BLACK MISCHIEF

BLACK MISCHIEF

Language, Life, Logic, Luck

SECOND EDITION

David Berlinski

HARCOURT BRACE JOVANOVICH, PUBLISHERS
Boston San Diego New York

Printed in the United States of America

Library of Congress Cataloging-in-Publication Data
Berlinski, David.
Black mischief: language, life, logic, luck / David Berlinski.—2nd ed.
p. cm.
Bibliography: p.
Includes index.
ISBN 0-15-613063-7 (pbk.)
1. Science—Philosophy. 2. Physics—Philosophy. I. Title.
Q175.B469 1988 87-22695
501—dc19 CIP

Second edition
A B C D E

For
My Parents

To
Victoria

Contents

PART II

PART III

Acknowledgements

I am grateful to John Casti and Hugh Miser for cooking up the deal that allowed me to spend time at leisure in Vienna at the International Institute for Applied Systems Analysis; and to René Thom, for making it possible for me to enjoy the hospitality of the Institut des Hautes Études Scientifiques in Paris. These are wonderful institutes and cities in which to work, and wonderful friends to have had.

Carolyn Artin and Klaus Peters at Harcourt Brace Jovanovich read this revised version of my book with great care, and reported their occasional reservations to me *sotto voce*, with infinite and endearing tact.

Preface to the Second Edition

I have taken the occasion of the second edition of this book to slash away at a number of passages, paragraphs, and pages that seemed to me to say twice what needed only to be said once.

The original version of *Black Mischief* contained neither a table of contents nor a set of chapters organized roughly by content: I had hoped that the thing might be read from front to back, or back to front, like the *Talmud*. My readers advised me simply to knock it off: They wished, they said, a table of contents and chapters that made sense. In this edition, they have both.

And as long as I am writing in a wonderfully forthcoming mood, I may as well mention a few other matters.

Any form of self-expression is necessarily a form of self-invention. The incidents that I record in this book are all of them true: Everything happened as I say it did. Yet the reader might remember from time to time that the character who appears in these pages, using my name, and apparently living my life as well, is not strictly speaking identical to his author, however much they may have shared a number of unforgettable experiences.

This book will not take the casual reader to the cutting edge of research. Nor is it meant to. What I am after in *Black Mischief* is the moment in which various lines in an intellectual field of force collect themselves into a kind of dense knot. Thus when I think of linguistics, the time is 1970 or thereabouts; Noam Chomsky is attacking B.F.

Skinner; I am a professor at Stanford University; I am twenty-eight years old. No doubt, a great deal has happened since then; and yet, I would argue, this was the turning point, the moment at which a clotted singularity appeared in the history of thought.

A number of otherwise sympathetic reviewers have suggested that my real aim in *Black Mischief* was somehow to show the persistence of certain outmoded Newtonian forms of thought in economics, or psychology, or biology, or wherever. Not so. My intention has been to explore a tangle of connected concepts. The idea of a Newtonian system is one such concept, but I discuss in this book microeconomics and utility theory, stock-market forecasting, random walks or romps, systems analysis, behavioral psychology, generative grammar, finite state automata, the Hilbert Program, artificial intelligence, differential equations, formal systems, theories of information and complexity, thermodynamics, entropy, the theory of probability, statistics, mathematical model theory, logic, the incompleteness of arithmetic, the theory of the singularities of smooth maps, and the neo-Darwinian theory of evolution. No one concept ties these separate concerns together. My book is unavoidably episodic. Yet the careful reader will discover that these separate subjects are unified by a number of themes and threads: I have tried wherever possible to organize what I have written in terms of the metaphor of a machine – the most general of devices taking inputs to outputs by means of a set of states. And I have stressed the crucial role in a certain class of arguments of the concept of a classification.

Finally, I have restored to a position of prominence certain sections of my manuscript that my hard-cover publishers assured me I could not print and they would not publish. In this regard, I will always treasure the memory of my editor standing beside his desk, an admonitory finger poised by the side of his nose, realizing, just as he was about to launch into a lecture, that the pages of my book he had thoroughly intermingled with the pages of a book documenting the life of Tina Turner.

David Berlinski

PART I

True Confessions

I have never been particularly eager to know how it is that the universe was formed, or how a magnet works, or why, for that matter, water flows downhill. I am comfortable with the thesis that objects unsupported fall toward the center of the earth because they have an affinity for the ground, and believe it superior to that offered by Newton. There it is – a certain implacable lack of physical curiosity. My interest in science is science: I am by nature a second-story man.

The great scientific culture of the western world – *our* scientific culture – achieved its first and most spectacular success in mathematical physics; for many years, it was assumed by scientists elsewhere that it was after physics that the other disciplines were intended by divine grace to gallop. This lent to the sciences as a whole an engagingly cinematic cast, with theoretical physics receding rapidly into the distance, pursued by any number of steaming subjects, their sides heaving and their flanks wet. Physicists still see the world in such vibrant Technicolor terms. I like the setting but not the shot, and incline toward a more up-to-date biological image, with science cast as one of those ominous protozoan-like blobs that Japanese fishermen were forever dredging into their nets in movies of the mid-1960s. With no apparent organizing center, and lacking even a head, the thing would divide, and then divide again: physics, chemistry, biology, economics, psychology, sociology, at which point screams commenced.

Every language, it is sometimes said, creates a special world. The universe of the fish-smelling, raven-haired Eskimo, much occupied with sorting snow into very many grades and shades (smooth snow, shiny snow, sad snow . . .), is inaccessible to me. In the contemplation of this circumstance, I experience no sense of loss. What holds for language holds for life as well. Without too much by way of intellectual discombobulation, I can imagine myself living as a Frenchman, or even as a bat, and worried thus about my liver or my ears. A style of the imagination in which the concepts of *accuracy, inference, theory, experiment, data, computation,* and *evidence* did not figure conspicuously, I would regard as pretty much one with the Eskimo.

And not only me. Were one to collect in a very large box all of humanity's scientists, from the first Babylonian, peering out at the night sky, to the youngest and brightest of contemporary physicists, talking credulously of the Big Boom in the brisk, inescapable present, the majority would yet be alive, and squirming, no doubt, to get out. This is one of those curious, hopelessly irrelevant facts forever washing over us like so many sandy waves. Indeed, the sheer volume of current scientific work is often taken as *evidence* for the vibrancy of scientific culture – on the assumption, evidently, that in order to measure the height of the Tower of Babel, it is necessary to count the number of languages spoken at its base. At first cut, it is true, there *do* appear to be any number of alternatives to the scientific system of belief. I have seen a volume on the shelves of my local bookstore in which the author undertakes the resuscitation of alchemy. There are other texts that treat of Palmistry (don't look), and ESP (don't listen), Tarot Cards, Witchcraft, Telekinesis, Astrology, Sun Worship, Astral Planes, Sun Signs, Faith Healing, Herbal Medicine, Nostradamus, Life After Death, Reincarnation, and the Bermuda Triangle – this in addition to the usual stuff on Getting Well and Feeling Better, Losing a Lover (or Loving a Loser), Couples, Stages of Life, Dressing for Success, Power Lifting, and Power Lunches. It is very disappointing to realize that each is in some sense a pseudo-*science;* even astrology, for which I have a tender, wanton sense of sympathy, is cast nowadays in computer-drawn horoscopes to ensure what a book jacket describes as "incredible accuracy." An acquaintance once urged upon me the thesis that *his* spiritual adviser was capable of levitation. "Here's *proof*," he felt compelled to add, thrusting a grimy photograph into my unenthusiastic hands. There in black and white, I could see rather a pudgy Oriental caught by the camera as he endeavored to hop from a sitting position, a thin wedge of light visible beneath his plump posterior.

Except for mathematics, a queer, remote outsider, dressed forever in earmuffs and muffler, even in high summer, the various sciences are involved in the search for purely a *physical* explanation of the flux and fleen of things. Like the pursuit of the Yeti, it is an activity with notoriously slippery standards of success. Certain elementary particles, physicists report, simply have no mass and exist in a state of anorexigenic purity. The neutrino is an example. Or used to be. That charming and old-fashioned high-school electron, trustfully circling the interior of an atom, has somehow been replaced by a vaseline-like smear of probability. Where *it* is, no one knows. Even space and time have long been merged into an inseparable viscous medium, something like Jell-O perhaps, or an unpleasantly undulating farina. Within molecular biology – I am passing from the cosmic to the comic side of things – ideological affairs are as they used to be in physics, but molecular biologists find it impossible to account for even the lowly bacterial cell without invoking items such as *information, complexity, order* and *organization*. These are concepts of stirring confusion and, in any case, not obviously physical in their cast.

For the hedonist or the haberdasher, the world is a world of matter. What you see is what you get. Now it is possible, I suppose, that a scientist might reject materialism in his spiritual life while accepting it in practice. On receiving word of his Nobel award, Abdus Salam (the Great Abdus) thought first piously to praise Allah instead of thanking the Swedish Academy. This suggests a world view in which the laws of matter run so far and no further:

Beyond is the realm of Spirit or the fearful Norse God of the Woods, clumping ever onward, solitary and remote. For the rest of us, a world of matter is the world that matters. What else is there?

The superiority of physics to subjects such as interpersonal communications or whole-body massage rests ultimately on the assumption that in explaining how the material world works physicists are in a position to explain everything else as well. Such is atomism, an ancient Democritean doctrine that has simply forgotten to disappear. The physicist thus tends to think of Nature as pretty much like an elevator: In going down, at some point one should finally hit bottom with a little lurch. This is the arena of the elementary particles, of which there are a great many. For all I know, looking at these things from the outside, the elementary particles, like the real line, might simply be infinitely divisible, with new particles popping up precisely in proportion to the funding available to investigate them. This is an awful thought. But even if physicists discover that in the end everything is made of quarks or strings or translucent space-time curves, no one believes any longer that the *principles* that govern the elementary particles (*all is flux* or *everything is fire* as appropriate as any, as far as I can tell) will ever explain completely the origins of life (luck) or the decline and fall of the Roman Empire (carelessness) or the fact that most men are born to suffer (fate).

From the dog's point of view, the world, I suppose, resembles largely a malign kennel deliberately arranged to thwart his pleasure – one damn rule after another: Do not urinate on the sidewalk, do not drink from the toilet, fetch, sit. No wonder the poor beasts slink so much and are so sneaky. Who am I to scoff? Experiencing turbulence at 39,000 feet, the cocktail nuts chattering in my anxious lap, I find it difficult to suppress the thought that Wotan is angry; my ordinary sophistication notwithstanding, I manage surreptitiously to offer Him a few words by way of propitiation (down big Fellah, down). Such are the ineradicable promptings of what paleontologists refer to as the reptilian brain – a customer who manages to shove his way forward in my life with absolutely embarrassing urgency. When left to my own devices, I imagine the whole of the observable world taking shape in terms of fantastic desires, inhuman spirits, strange forces, with everything intentional, filled with marvelous or miraculous purpose, even the wheeling stars in the staring sky spelling out a systematic but incomprehensible message. At some level of analysis (*Genesis*, perhaps), this vision, I suspect, is true. It is, of course, not the level scouted by *The National Enquirer* – **Tough Priest Exorcises Vicious Demon from Pet Pig.**

Science, at its most successful, sets a stern face against the ancient doggish desire to interpret physical events in terms of the categories of intention and belief. It is thus that *mechanism* has come to replace animism in contemporary thought. This is a historical development in which the physicist or biologist takes inordinate and preposterous pride, rather like some mournful movie monster admiring for an altogether perishable moment the steel plate stapled to his own skull (so nice, so sturdy). In a narrow sense, a mechanical explanation is one couched within the terms of Newtonian mechanics – a scheme that is unable to encompass electromagnetism or thermodynamics, and that appears false at the margins of our experience in that bizarre limit where things are very large, or very small, or very fast, or very far. Curiously enough, in the concept of a machine, mechanism has itself outlived mechanics – satisfying evidence that in science, unlike life, nothing is ever given up for good or lost forever.

Getting Rich

A lecture at the University of Washington having been arranged, I took the night train from San Francisco to Seattle. I ate in the dining car, still plush in the mid-sixties, with heavy napery, curiously dense silverware cut from stainless steel, and pleasant moon-faced waiters, who rocked in the aisle and seemed able to pour steaming coffee from a variety of improbable angles. Afterward I retired to the coffin of a sleeperette. At the station I had picked up a paperback with the appealing title *How to Make a Million Dollars*. The author was, I seem to remember, a Canadian orthodontist, whose self-satisfied face peeped from the back cover, a sly smile across the lips. I've been poor, the face seemed to say, and I've been rich, and, my friends, believe me, rich is better. The author's secret for financial success seemed absurdly simple. One bought when things were cheap and sold when they were dear; or vice versa in the case of an operation called short-selling, which at the time I confused with stopping short, and thereafter imagined as the fiduciary equivalent of a rear-end collision. Investing in commodities, one might make a fortune in days, even hours, occasionally minutes. There were options, puts, calls, annuities, convertible debentures – convertibles of some sort, a kind of Cadillac of securities – preferred offerings, hedges, warrants, hot tips – financial instruments of such melting pliability and outrageous availability that a man who failed to make pots and pots of money after giving his investments only the briefest attention need actually have approached the matter determined to lose, and was probably in search of some form of public abasement, like a madman or a monk.

At the department of philosophy's scruffy third-floor office, two desks, several green metal chairs, a wooden pigeonhole for letters, and a hissing radiator had entered into an inhospitable conspiracy. Waiting for me in an embarrassed circle were the chairman, whose bald head rose conically from a turtleneck sweater like some majestic growth, a shaggy specialist in Oriental philosophy with a full beard, hairy ears, and the shambling air of a man consecrated to a language he cannot master, and two or three younger men dressed in chino slacks, heavy woolen work shirts, and boxlike boots. The secretary, who sat at her desk, hands chipmunked over her typewriter, had flaming pink ears, which functioned anatomically as a plaintive beacon calling in sailors – any sailors – from the sea. The chairman, our leader, suggested lunch at a Chinese restaurant. "Cheap," he said.

We drove for what seemed an hour and then parked in a narrow alley filled with the smell of grease and noodles. Inside the restaurant there was linoleum on the floor, and Formica-covered tables, with their inevitable neat arrangement of hexagonal bottles containing soy sauce, vinegar, and some blindingly hot red liquid that my host sloshed onto his plate, when it arrived, with a gusto suggesting a peculiar, rapt, incendiary passion. In the corner of the room stood a black-bordered fish tank in whose murky gloom swam several plaintive carp, eyeing the waiter and the fish net with an air of resignation and foreboding. Everyone ate rapidly, with the rather revolting heads-down posture that the Chinese adopt when alone, clicking their chopsticks as if they were castanets. Only the Orientalist demanded and then used a fork.

Afterward, the bill came. This prompted an inspired financial exercise. "Let's see now," said the chairman, suddenly alert, drawing a thin stream of air over his pursed lips to signify concentration, "there are five of us but I only had half the shrimp, so my total should be four eighteenths." The Orientalist excused himself to retire to the lavatory, and did not emerge until we were all in the street. I reached reluctantly for my wallet. "Absolutely not," said the chairman grandly, holding his hand up to stop me. "You're our guest. Why don't you just take care of the tip?"

The doctor was right. Rich is better.

The Dismal Science

Economics, it is often said, is the dismal science, but given the rivals already in the field – Black Studies (Did you know that Galileo was one thirty-second part black?), Ethnomusicology (Kurdish people are capable of whistling through their noses), or the Sociology of Couples and Significant Others – this last the speciality of a sociologist I once knew named Pepper – the judgment is close. Still, there must be something to the description; I find myself wandering intellectually whenever I look at an economics text. Since the publication of Gerard Debreu's *Theory of Value* in 1959, economics has become ever more mathematical, with results that are inscrutable even to economists. Great mathematicians such as Steven Smale now flaunt their skills defiantly at the economists by describing economic theory in terms of truly monstrous technicality. Literary economics – the description retains a touch of the derisive – seems to me, at least, to have become a branch of theology, supply-side economics, in particular, now cast in terms appropriate to the doctrine of the ubiquity of the Body of Christ. In any event, economists are as unhappy with economics as everyone else and begin their books with an excuse – *things could be worse* – and end them with an apology – *life is complex*.

Like a submarine sandwich, economics is double-layered. On the top, below the mayonnaise, there is macroeconomics, a subject in which various individual economic quantities (income, wages, taxes) are aggregated; on the bottom, above the provolone, there is microeconomics; here the individual and the firm hold sway. Theoretically, macroeconomics is designed to surge upward from microeconomics in the same elusive and bewildering sense that thermodynamics is often said to pop with a fizz and a smile from the surface of statistical mechanics. In fact, the great theoreticians of industrial life – Keynes, for example – treated income, wages, and taxes as if they marked the point at which economic analysis began. Microeconomics they looked at from above, with only the vaguest of interest. The formidable task of integrating micro- and macroeconomics is now very much the province of the econometricians, who figure in economic ritual as priests of a separate cult.

At the beginning, and the bottom (to give up those lunch counter metaphors), are land, labor, and capital, the ancient imprint on economic theory of economics as an agricultural treatise, and then the economic agent, who is presumed to be self-interested and acquisitive as a shark. Facing a world of scarce resources, he endeavors to get his share and keep it. Now every human action is the result of some secret affinity between what a man believes and what he wants. The chemistry of this copula economics cedes sensibly to psychology; what a man believes is, of course, his own business. His desires or economic wants – his *preferences* – contemporary economists regard as the raw stuff from which the theory of consumer demand is built.

Life being what it is, I may from time to time choose what I do not prefer, or prefer what I do not choose, but generally my preferences are reflected in my choices. This circumstance tends to force a man's economic desires to the surface of things. Preference itself economists describe as a relationship between an agent and what are often called *commodity bundles*. The analysis that follows is straightforward and simple. I may prefer one bundle to another, the second to the first, or be languidly indifferent between them and prepared to accept either if I am disposed to accept anything at all. My preferences may be represented by any mathematical measure in which the order of the numbers and the order of my preferences coincide. These numbers form a *utility* index, and are unique only up to a monotonic transformation – a utility index multiplied by a positive number is again a utility index.

In the tritest of trite circumstances an economic agent faces a world of choices. His problem is one simply of attending to what he wants; his goal, to *maximize* his satisfactions. Whatever he chooses he gets, in the sense that his choices are unaffected by chance. In explaining *why* this remarkably simple character when hungry exchanges his money for bread rather than books, the economist need only observe that the first is *preferred* to the second – presumably because paper is inedible. *You can't eat them books, man*. This lighthearted account may not appear to shed a hot white light on the subject of economic choice, but with the concept of preference in hand (or wherever), the economist can describe the behavior of an arbitrary economic agent – this by means of a unique set of demand functions. It is difficult to improve on the analysis that results, and a source of happiness among economists that having said this much about preference, one hardly need say any more.

Nineteenth-century economists, on the other hand (and keeping rhetorical forelimbs to a pair), assumed that the economic objects that an economic agent wanted, he wanted in virtue of their *utility*, where the utility of most items (wives, for example) was as palpable as their weight, and thus a quality with a robust and uniquely measurable numerical structure. The utility of a set of similar objects (wives, again) the economist represented by a curve that is at once monotonically increasing but negatively accelerated. Human beings are infinitely acquisitive (that curve rises forever) but easily bored (it tends over time impotently to droop). Such is the principle of diminishing marginal utility, one of the great tragic declarations of the human race.

The idea that the utility of an object is a property rather like its color serves consolingly to unify a world of value and a world of fact. In our own century, we have come to insist that values are *not* part of the panoply of properties that in some primitive sense are simply Out There. In any event, measurable utilities came to grief quite on their own. For one thing, no economist could quite make sense of the interpersonal comparison of utilities – one man's satisfaction compared to another's; yet if utilities were palpable as weight or color, some judgments between economic agents should surely have been possible. Apparently not. When Vilfredo Pareto discovered that the technical results of utility theory, chiefly the substitution and income effects, could be generated by indifference curves, a pleasant conceptual economy was the result.

The natural thought that money should be the standard by which the utility of various commodities is appraised led to difficulties in the analysis of insurance and gambling. Many nineteenth-century economists believed that no sensible man would risk a dollar to gain a dollar in a fair gamble. They reached this unpersuasive conclusion by meditating copiously on the doctrine of diminishing marginal utility. Alfred Marshall, for example, wrote in the mathematical appendix to his *Principles of Economics* that "the argument that fair gambling is an economic blunder . . . requires no further assumptions than that firstly the pleasure of gambling may be neglected; and secondly . . ." – and here I translate into the vernacular – that the curve measuring the utility of wealth slopes downward as wealth itself increases. Marshall found himself so wobbled by the juxtaposition of these assumptions that he sought solace in the doctrine that risky decisions do not involve the maximization of utility at all. His successors – Edgeworth, Fisher, Pareto – looked on this with amazed contempt. Marshall, they argued would have been better off had he simply chucked the whole idea of diminishing marginal utility, chucked, in fact, the concept of measurable utility altogether.

In the raw world of real choices ("raw" and "real" being adjectives with an uncommon affinity for one another), time and chance have an unhappy way of interposing themselves between what an economic agent wants and what he gets. The result is economic choice under conditions of risk (the odds are known) or uncertainty (who knows?). Here, economic theory merges fluidly into the more comprehensive theory of rational decision-making itself, where choices need not be economic: Beyond are various statistical theories, and even the theory of games. In the analysis of any risky business, the economist treats the contingencies confronting an economic agent as *states of nature* (a wonderfully Hobbesian phrase that), with Nature now cast as a personal antagonist capable of appearing in a finite number of guises (a courtier, a courtesan): It rains: It doesn't rain; The plane crashes: The plane lands safely; War: Peace; Life: Death. Examples in this domain are notably unimaginative. These states of nature are crossed against the choices that an economic agent might make, so that each choice is associated with a finite set of possible *outcomes*. Given some assignment of probability to each possibility (by the weatherperson, say), the overall utility of a particular choice appears now as a cross between the probability and the utility of the outcome – its *expected value*, what one most wants discounted by the likelihood of obtaining it. But in computing expected values, or almost anthing else of real interest, the economist requires utility numbers that reflect preferences to a specific numerical degree, as when one wife is not only better than another, but better by a fixed and unalterable amount.

Preference is an austere but attractive concept. It allows the economist, in talking of what an economic agent wants, to point simply to what he does. This affords him the satisfaction of imagining that economics is a behavioral science. Economists and psychologists have the warmest feeling for one another, and often arrange that their offices be side by side. But microeconomic theory is sufficient only for economic choices made under conditions of certainty. Elsewhere, measurable utilities make an ectoplasmic reappearance – this to the embarrassment of economists, who thought that they had put all that behind them.

Losing Money

Beyond the sexual, all human beings may be divided into two classes: There are those to whom money flows and those from whom it recedes. I have always found myself squarely in the second camp, surrounded by others of life's fiscal losers; but once a wife that I subsequently misplaced foolishly turned over her savings to me, and I found I had more money to spend than I needed to live.

Having immersed myself in matters of finance by mastering the contents of one slim paperback volume, I turned eagerly to the stock market. My first broker I now recall as having a face of such unenterprising regularity that it seems to have been enthusiastically adopted in succession by any number of distinct characters interested in anonymity: An airline steward on a flight to Copenhagen, a used-car salesman in Palo Alto, a grunting Kung Fu instructor in Seattle, my daughter's escort to a concert given by some hideous transvestite.

Of my broker's name I remember only that it began with a soft chuffing sound and ended in a consonantal clunk: Chuck, perhaps. His office consisted of a cubicle whose walls were made of some thin translucent plastic, one of a series of identical offices off a long corridor. As I entered and identified myself, Chuck rose in a tigerish crouch, took my hand and crushed it, and squeezed my shoulder in a gesture of conspiratorial friendliness. "Hold my calls," he said grandly into the telephone, after dialing what I imagined to be his secretary. "I'm in conference."

I had come to the brokerage house as the result of my broker's cold call, a device of salesmanship wherein the young broker contacts the first three hundred names from a source so standard as the telephone book, and, after suffering silently any number of peremptory rejections, which, I subsequently learned after marrying a broker of my own, ranged from the terse (*No!*) to the trenchant (*No!*), endeavors to communicate to those that remain the inestimable advantages to be accrued from a financial consultation in the near term.

Looking up at me from gray eyes that were tipped with sable lashes, an elegant combination, really, he said: "So you're a prof, eh? What do you do for a living?" and, after hearing from me what I really did, responded amiably that it sounded better than work.

In explaining his rosy investment strategies to me, Chuck thrust his face next to mine, as if to prevent eavesdroppers from learning of the secrets whereby directly I would be rich. In his attitude toward the market, my broker was a fundamentalist, which is to say he believed in the existence of the true value of a stock and sought for opportunities to cash in on the difference between a stock's price and its value. He had conceived a passionate attachment to a firm called Blatstein-Goldhaber-Ventriss, admiring from a distance its by now mythical founder, a character, I later learned, whose chief financial gift appeared to lie in his miraculous ability to leave office before the process servers arrived, and enthusiastically endorsing the *concept* of the firm, which, as far as I could understand such things, involved the ingestion of one corporation financed entirely from the digestion of another.

I agreed to purchase BGV warrants at 53, and left the office squinting at the bright sunshine of San Francisco, convinced that my broker's muscular hand upon my shoulder signified nothing less than the palpation of my financial future.

For a week I enjoyed the ineffably delicious feeling a man acquires on knowing that presently he is to become very wealthy. Soon enough, the roseate peeled-peach color of my dreams turned distinctly darker. My wayward warrants from BGV ascended slowly to an agonizing apogee at 54, and promptly turned downward. "Just a technical correction," Chuck-the-Cheerful advised. Gathering speed, those warrants hurtled past a panorama of points marking what Chuck-the-Chastened now called *levels of resistance* (there were many levels, not much resistance) until, finally, they descended into the lower depths of prices in the high twenties, whence they had risen some six months before.

My second broker – Hans or Horst – was a vague European who grimaced with annoyance when I noticed his distinctive *Ch*erman accent. He wore iridescent silk suits, and shirts with collars so high that when he sat their back edge disappeared in the looping fringe of hair that curled over his neck. He heard out my witless tale of warrants while drumming his fingers on the table and stealing glances from my face to look at the moving ticker-tape above my head. "*Nah, zo,*" he said sibilantly.

It was his belief that investment involved nothing so much as a series of quick nipping movements into the market, an operation that I pictured in terms of goldfish feeding. My talk of the real as opposed to the present value of a stock he dismissed impatiently, as indeed he dismissed all forms of precise knowledge of economic theory – this on what seemed to me the sensible view that knowing too much could actually confuse a man. "All the information I need is here," he said, stabbing his finger at a complex point and figure chart. The historical record of prices already achieved provides the surest and the only indication of prices to come. "Trust the market," he said reverently.

Hans-or-Horst followed Occidental Petroleum and endeavored to so time his purchases as to take advantage of minute fluctuations that other brokers would find imperceptible. There on the chart he pointed out a womanish Head-and-Shoulders configuration and observed sagely that, like a trusting seal, Occidental Petroleum was swimming slowly between two technically significant trading points, which in my innocence I took to correspond to the chart's nipple and her nose.

It did not occur to me, as he was outlining all this, that he was explaining a strategy for *his* financial success, and not mine.

On another occasion, I determined to enrich myself in the commodity market. I ordered and then read several mimeographed publications dealing with cocoa, sugar, wheat, and frozen pork bellies. The specialist to whom I presented myself at E.F. Hutton, unlike the stockbrokers I had known, treated my transactions with the natural wariness of a man inspecting a large and sinister snake. On the strength of nothing more than a hunch, I had suggested buying sugar long. "Gonna get burnt there Dave," he said. "Lot of potential for downside risk. Let's try shorting bellies." Frozen pork bellies zigged when they should have zagged; anticipating delivery of a contract of oleaginous bacon, I bought at a loss. Sugar, to which I had turned a blind eye, ascended majestically to the limit each and every day for a period of six weeks. I was thus able not only to compute my losses but to compare them simultaneously with the gains I might have reaped. My friend, the mathematician Daniel Gallin, on observing the pleasant mornings I spent visiting my brokers and stung, no doubt, by the thought that soon I would be rich while he would yet be poor, decided on the advice given him by a successful cousin in Los Angeles, to skip the financial markets entirely, and sank his life savings into a feature-length film about a pornographic pig.

The Axis of Desire and Belief

Within the sweaty economic arena, an economic agent confronts the entrepreneur or firm, his natural antagonist and *doppelgänger*. It is the entrepreneur who would sell what the individual might buy, and who thus produces what the consumer consumes. Individuals and entrepreneurs in microeconomic theory share a certain secret symmetry of structure. In appraising various commodities, the individual has recourse to his utility index: He acts to maximize his satisfaction. In casting his bread, or anything else, upon the waters, the entrepreneur has recourse to a *production function*, an imaginary instrument by which he balances the various costs of production (labor, capital, affirmative action) to best effect: His aim is to maximize his profits. The actions of individuals and entrepreneurs alike thus fall under the same mathematical concept: maximization or optimization.

An individual wants, but individuals *demand*. A firm produces (whips and thongs, say), but firms together *supply*. It is thus that two great economic forces are created. Together individuals and firms act to form a market. Here supply and demand are aggregated, or heaped together. Under circumstances that are ideal and imaginary, a market is properly described by the theory of perfect competition. It is a theory with a simple, trusting, geometric heart. Two curves are paramount: the demand curve and the supply curve. In the first, consumer demand is pegged against price on a coordinate axis. The cheaper the price, the more consumers demand. Demand curves thus slope downward in a melancholy way. In the second, it is supply that is pegged to price. The more expensive an item, the more suppliers will tend to provide. Supply curves ascend solemnly. The curves taken together represent the axis on which economic theory ultimately revolves. Where they intersect, an equilibrium is reached. Supply and demand precisely balance one another, and exist in a state of perfect equipoise.

When left to themselves, individuals and firms in a free market move toward equilibrium by means of a series of linked trades. Unlike contracts in the law, economic contracts are made to be broken. Individuals and firms in this particular world are free to recontract for the purchase or sale of a commodity the moment that the price breaks in their favor. For many years economists simply *assumed* that this sort of amoral but aimless activity would end inevitably in equilibrium. Not so, as it turns out. Agents in a perfectively competitive market may fail to resolve their differences at a single point.

The theory of perfect competition is elegant, and makes uses of no mathematical concept more complex than those found in the calculus. It may be taught easily to undergraduates – reason enough, economists believe (or say), for the thing to be cherised regardless of its truth. In a world of sunny doctrines and shady deals – *our* world – prices are plainly *not* set by anything like perfect competition. But microeconomic theory and the facts part company at an even earlier stage. To demonstrate that an ideal market will reach an equilibrium, economists must assume that the various traded commodities are homogeneous, with co-ops and condoms treated as if they were drawn from a common stock. That condominium purchased in April may stand empty and unsold in December, simply because the condominium market is too thin to sustain competition. In the warm glow of theory, there are *always* numerous buyers and sellers in the market. Out on the street, it is cold as the mouth of Hell. Participants in a perfect market have perfect information about their trades – I am continuing now with a list of counterfactual assumptions; in almost any real market, information is itself a valuable commodity, and reaches the various actors at wildly uneven speeds, with the losers and the luckless often the last to know. A perfect market is open and unconstrained. There are no forced purchases, and uniformly distributed transaction costs, which, like country-club dues, participants share without a peep of protest.

Still, it would be uncharitable (and pointless) to make too much of all this. A theory may be interesting, though false, and trivial, though true. Economists, no less than mathematicians, are entitled to think of certain theoretical constructions (a perfect market, the derivative of a function) as ideal points – limits, in effect, toward which things or processes in the real world may be tending.

A Prince of the Realm

Toward the affairs of life, Aaron Wildavsky conducted himself with casual and endearing inperiousness. Sometime in the seventies I had sent him a copy of a little paper that I had written; he responded with warm effusiveness. I concluded, naturally enough, that his intellect was a thing of great and inescapable grandeur. When I visited him at Berkeley I was surprised to find that he resembled a longshoreman or a long-distance truck driver. He was bald, bearded, fierce-faced, and spoke with a charmingly thuggish New York accent.

Now he was on the telephone. He was in line, it seemed, for a National Science Foundation grant best measured in terms of the fiscal equivalent to a light-year. I listened alertly.

"Dese guys," he began – our Aaron, Dese guys! – "want something on technology assessment."

I was in favor of the assessment of technology, the technology of assessment, whatever.

"We need something in about six weeks," Aaron said, and then outlined the details.

"Innerested, fellah?"

Innerested.

In due course, I found myself a member of a tripartite NSF team. Wildavsky was our leader. The team's other member was a Ph.D. candidate in political science, an immensely slow, thick-witted creature, heavy in the haunches, with a dense grape-black beard, and an intolerable, even maddening way of drawldroning his way through any act of communication. Every Tuesday I would fly to Berkeley from Seattle for a meeting of the team. Heavy-Haunches would read slowly from his twelve pages of notes until Wildavsky and I were apoplectic, our irritation having absolutely no discernible effect on the pace or cadence of his speech. I would talk extravagantly of whatever came into my head, and Wildavsky would tug at his beard, and pull at his ears, and run a brown hand over the bronzed dome of his head, and snort derisively.

Hot Air Rising

Technology assessment afforded me my first and only opportunity to apply economic principles to the real world, a circumstance by which I was persuaded that I was genetically indisposed toward the solution of any practical problem whatsoever.

There was air pollution, for example. On many days, Los Angeles is largely engulfed by an acrid mist that from a sufficient distance – 35,000 feet straight up, say – looks like poorly made lentil soup. Such is photochemical smog. Its cause and chemistry are by now well known. Certain peculiarities in the shape of the Los Angeles basin make it impossible for hot air to rise. Automobile exhausts, like certain producers, having no place to go, stay close to the ground. There they are catalyzed by the remorseless southern sun, which, midway between the first and second casting calls, fixes a chemical pathway leading from hydrocarbons and nitric acid to ozone and nitrogen dioxide. As the sun reaches its zenith, the smog ripens like an avocado, and the sky, which at six in the morning looked pale and pink, turns raw-rose red and yellow. The steady gush of hydrocarbons falls during the evening hours; sea breezes stir and some of the yellow is washed out of the air and replaced with striking and exotic violets and blues – the detritus of photochemical smog. Every day, the cycle repeats itself, like a ritual, except for those rare sparkling days when a stiff wind blows in from the ocean and sweeps the city clean.

For the longest time, smog was considered a natural affliction, something that just happens, like short legs. The discovery in 1951 that photochemical smog was actually an automotive pollutant had the effect of a very slow-moving thunderclap. Within twenty-five years, technologist and politicians managed to complete the obvious chain of inference: In order to cleanse the atmosphere, it would be necessary to control automotive exhausts. The Clean Air Acts, and their various amendments, owe much to this insight. But despite the fact that many automobiles now carry catalytic converters, the ambient air above Los Angeles has remained much as it always was, which is to say, dirty.

The rising sun and the internal-combustion engine, the chemist Julian Heicklen observed, represent inescapable absolutes. A catalytic converter may change the character of automobile exhausts, but not by much. Still, a chemical pathway that cannot effectively be blocked at its source may well be interrupted at some later point, in the air and not on the ground. What Heicklen specifically had in mind were free-radical scavengers. Added to the air in much the same way as crystals of dry ice are scattered over clouds, such scavengers, so the theory ran, would block the final furious chemical reaction that produces ozone and nitrogen dioxide. This would not purge the atmosphere of hydrocarbons, but if Heicklen's proposals were sound, and they seemed to involve nothing more than elementary inorganic chemistry, the upward-moving eye would see only a black smudge on the horizon instead of the usual yellow mist.

Heicklen's proposals generated only dim incomprehension among his colleagues. Officials at the National Science Foundation, bewildered by their first encounter with a new idea, took pains to see that Heicklen's ideas were evaluated by his academic enemies.

Those enemies promptly turned their attention to diethylhydroxylamine (DEHA), which appears as a by-product of the chemical reaction that occurs when the pathway to photochemical smog is blocked. They suggested that DEHA might well be as dangerous as photochemical smog itself. Heicklen and his associates Carl Meager and James Weaver conducted standard animal toxicity tests on DEHA, and reported themselves satisfied that DEHA was not a carcinogen. Heicklen thus argued that DEHA was safe because it was not obviously dangerous. James Pitt, a chemist at the University of California, argued some months later that, in fact, DEHA shares certain structural properties with a well-known carcinogen – diethylnitrosamine. Pitts thus argued that DEHA was dangerous because it was not obviously safe. From this concussive clash, I was supposed to draw some conclusion.

Fat chance.

Taking Stock

In this country, but not in Cuba, Albania, or the Soviet Union, an individual may earn his living by selling just about anything from soap to sushi (much the same thing, as far as I can tell). If he profits, *tant mieux*. He gets to keep the rewards of his toil, or to reinvest them in business. If he flops, *tant pis*. Such are the constraints of capitalism. Now by their nature, as Lucretius might have put it, economic entities (or agents), like tumors, endeaver to grow. In constructing a larger enterprise from a smaller one, the entrepreneur acting alone faces certain obstacles. There is a limit to what the banks will lend the solitary operator talking dreamily of solar windmills, brass nose rings, or soap-on-a-rope in the shape of a phallus. Then, too, one man can do only so much. It is for these reasons that partnerships exist. In a partnership entrepreneurs come together, as in a marriage, sharing their risks and their rewards, yet retaining somehow their individuality as well. Partners to a firm may be sued separately, for example. The man who enters into an association with his brother-in-law is liable for his debts when Morty, Mo, or Max, acquiescing finally in a sun-splashed criminal impulse, departs for the Cayman Islands or Brazil, the partnership monies only a cocaine smudge around his nostrils and upon his upper lip. With such contingencies in mind, lawyers make admission to partnership an arduous ritual. By the time an associate gets to use the executive lavatory, he is simply too tired to consider absconding with the firm's assets or raping its secretaries.

Partnerships distribute risks, but not very widely; they make the accumulation of capital possible, but not to any great degree. These limitations a corporation transcends. In the eyes of the law a corporation is a legal entity – a *person*, in fact, with all the legal rights and powers of personhood. As a legal entity, the thing is entitled to raise money, participate in commerce, and absorb those vagrant liabilities and risks that would otherwise devolve upon its all-too-human constituents.

Unlike an individual or a partnership, a corporation

raises money by selling aspects of itself, a process that is theoretically infinite, and that suggests, somewhat paradoxically, that corporations are infinitely divisible. Such are the stocks by which ownership of a corporation passes into public hands. On the view of things patiently projected by high school civics textbooks or brokerage house brochures (the doubled analogues, in point of dullness, to Trenton and Tacoma), the ultimate authority over a corporation is wielded by its stockholders, with management piously in their service. The real history of the corporation has been one in which, like a Gabor marriage, ownership and management have grown progressively estranged.

A corporation exists in theory as a fictitious person; it acts in fact as an agent constrained by only a single desire, which is to maximize its profits, that wedge between what it spends and what it gets. From time to time, the profit motive comes to appear morally hoggish, even to the men running the major corporations. Much effort is devoted to suggesting somehow that corporations exist for more socially redeeming purposes than raking in the loot. While living in the bleak northwest (to describe the place twice), I was especially taken by an advertisement in which the themes of selflessness and a commitment to the environment were suggested on television by a trailing shot of a dwarfish Indian on horseback. The whole thing was a marvel of corporate concern. Much exercised by litter, the imperturbably woodsy Tonto managed to force a tear to trickle from his canthus to the point of his mahogany chin. Through one of those queer, unanticipated twists of history (what Marxists refer to as the cunning of the Dialectic), the profit motive has reacquired its satiny moral sheen: It appears in popular consciousness as a corporate adjunct to the injunction to be all that you can be. Were the commercial to be reshot, no one would think to focus on the environment. An indefatiguable blonde in leg warmers would be advising the rest of us to firm it up, shake it out, and grab it all. The music on the soundtrack would describe a hammer beat of irritation.

And that Indian would have to go, of course.

Bi-Coastal: East

For the congenital cutup, academic life is an infinitely descending ladder. I had taught at Princeton and at sunny Stanford, which smelled of eucalyptus leaves in spring, and which I left as the result of my mistaken impression that I had something better to do with my life than commute to work along the Bayshore Freeway. I had been unemployed for what seemed a very long time – an experience rather like that predicted by the general theory of relativity, in which time expands while the work available to fill it contracts. One evening the discreet and serviceable white Princess telephone by my bedside chirped into agitated life; from the employed beyond, a neat, reserved, self-contained voice identified itself – "Peter Kivy here" – and asked courteously, with even a delicate touch of diffidence, whether I would be available for an interview with members of the department of philosophy at Rutgers University.

It was thus through the action of an autobiographical jump cut, in which the tail of one disaster is affixed to the head of another, that I found myself engaged again in academic life.

My routine was always the same. Low and sleek, the Metroliner from New York would pause at the Newark station, exhaling mist into the cold air, like an athlete gently panting. I would bury my face into the collar of my imitation fur coat and walk down the steps, past the graffiti-covered walls with their inscrutable messages – **Shabu '79, The Silver Fox, ASPOX** – and into the chilly streets of central Newark. Moving northward past the gloomy garbage in piles, I passed a row of grim two-story houses, and a liquor store, and a store in which a gold-toothed Cuban emigré was fashioning panatelas, and a vacant lot, where a group of men were warming their hands over a trash-can fire and saying "Sheeyit" to one another with immense conviction.

I had expected that my classes would be composed of sullen blacks, eager to offer me an act of racial revenge, and glimpsed in tabloid my anticipated future – **White Prof Stabbed By Eleven-Foot Black** – and my assailant's expected defense – *He was fuckin' over my mind, man* – but my students were New Jersey ethnics chiefly, and the blacks who came to my lectures treated me with sad, furry softness. My class list stretched from Agnostino to Zfez, but of the three hundred or so students that I taught over the course of three years, I remember only Lisa Dumbrowski, who chewed on her inky pinktails; Raymond Svic, who aspired to dentistry (not in my mouth, you don't); Rufus "John F. Kennedy" Jones, who answered to his name with a bell-like "Yo"; Boom-Boom Salacio, of the Sicilian Salacios; and a delicious young woman with straight hair and blue eyes, who would enter the room late and walk to her seat from the stairs, her lavish hips swaying gently. Analytic philosophy, which I taught with what seems in retrospect to be a flamboyant lack of skill, appeared to my students as yet another impediment artfully designed to irritate without enlightening. After struggling, with marked good nature, to make sense of Descartes, or Berkeley, or Hume, they finally elected to treat the entire subject with baffled but respectful indifference. At the buzzer marking the end of class, they would rise in a group like some multi-pawed mammal, troop toward the rear door, and disappear.

And my colleagues? There was an English statistician, as I recall, with a very long nose, sandy hair, and pale, washed-out eyes. I found his company congenial if only because he acknowledged with perfect candor that teaching regression analysis to undergraduates was a professional activity of complete and perfect worthlessness. "The little buggers can't even spell their own names," he once remarked as he was marking a computer printout in blue pencil. His own secret passion was theoretical physics. He was engaged, he told me, in a far-reaching program to acquaint himself with contemporary developments in the theory of elementary particles. A follower of Arthur Eddington, he believed that Eddington had been on the right track in postulating some fantastic space of six or six hundred dimensions in order to account for both general relativity and quantum mechanics. That no other living physicist had thought to follow Eddington on this particular road, he took to be a sign of their inescapable hostility to genius. At each of our meetings, he would part with a larger and larger portion of his exotic vision, indicating to me with mounting excitement and a queer mad look in his pale eyes that he was very close now to being able to formulate the single previously unutterable principle by which all other principles might be subsumed.

The chairman of our little department – *unser Führer* – was neat, fussy, hopelessly depressed; he wore a Mickey Mouse wristwatch and devoted himself to the philosophy of music, writing papers, I imagine (I never read any of them), on such subjects as the soul of the tuba. The department's logician was very much concerned with pragmatics, another dismal, doomed subject, and viewed his brusque rejection by the editors of various learned journals as evidence of a far-reaching spiderish conspiracy. When the editor of the only journal ever to publish his research results retired, it was regarded in our department as a desperately evil omen, something like seeing a stigmata appear on a dog's paw. Only Charles Biederman, the senior member of the department, who had a florid face with a large, aggressive, vein-marked nose, and who was tremendously over-weight, seemed actively to enjoy his life. Good old Charley! He spent his time, as I recall, trudging from one expensive New York restaurant to another. One day he succumbed to a coronary catastrophe, his fork poised in midair, midway between the pâté and the *pâtisserie* of a meal he never properly finished. At his sad, bleak funeral – the smell of decaying flowers in the air, three or four independent mourners, the widow, honking noisily in the front row, inconsolable – our chairman could think to say of the deceased only that he never stabbed anyone in the back. Small wonder, murmured another colleague: He couldn't reach that far.

On the vast concrete square that served as the grounds to the campus, where cripples in their hideous motorized wheelchairs scooted back and forth, it was always blazingly hot or miserably cold.

Whatever It's Worth

The laws of supply and demand hold for stocks as well as anything else. When demand exceeds supply, the price of a stock rises until supply catches up to demand like a second wave. With supply caught up, the price ceases to rise; supply and demand are in equilibrium. When supply exceeds demand, the price of a stock declines, until demand catches up to supply and a new point of equilibrium is reached. "There are only two things that can influence the price of any commodity in any market," Joseph Granville writes. "The first is supply; the second is demand. There is no third thing." To this extent, a market in securities is pretty much like any other market. In the world beyond the stock exchange, however, the *flow* of goods and services has a definite and irreversible direction. An automobile commences its existence on the drawing board, and after a number of years ends in the wrecker's yard, rusting and forlorn. To be used in economic life is to be used *up*. In the stock market, one and the same object may be acquired on numerous occasions. The man who sells IBM at thirty may buy it again at forty, convinced that it will ascend to fifty. For at least a time, stocks travel in circles, or concentric spirals, and seem as fresh and as valuable late in their life as they were the first pink moment at which they were issued. Stock markets may not have all that much by way of a classical structure after all.

Among other things, a stock is a legal claim against a corporation; theoretically, the value of a corporation should, like incident light, be faithfully reflected in the value of the stocks that it issues. On the world's stock markets, however, whether in New York or Tokyo, what are on their face mere legal instruments – double fictions, considering the corporation's status as an imaginary person – take on a vibrant, frantic life of their own, and like a headless chicken seem able for a time to suggest that they might carry on business all by themselves. In the end, of course, that missing fiscal head does assert itself; but why it is that stock market prices should display such astonishing short-term volatility remains a mystery.

On one way of looking at things, a corporation is valued in terms of the sum of its assets: fixed stock, equipment, receivables, goodwill. The stock that a corporation holds is itself considered an asset. The theory that a stock reflects the value of a corporation is in part, at least, offset by the observation that the value of a corporation reflects the price of its stocks. In performing this sort of numerical exercise (under just the right circumstances), the financial analyst may reach the conclusion that the outstanding shares to a corporation are worth more than the corporation itself. This is always an ominous sign.

In a charming and irreverent book entitled *A Random Walk Down Wall Street*, Burton Malkiel, an economist with a rare gift for well-tuned prose, begins at the beginning.

"What is it," he asks, "that determines the real or intrinsic value of a share (of stock)?" What indeed? According to the theory that Malkiel himself is prepared to dismiss with a sniff and a snort, the rational investor should purchase stocks with an eye fixed on their discounted future value, where the future value of a stock is measured in terms of the dividends that it yields. Those anticipated dividends are merely an index, of course, measuring more intangible items such as corporate growth rates, degrees of risk, market interest rates, or even price-earnings ratios. Regardless of a stock's actual price in the here and now of trading, its value is something else, a kind of shadowy secondary price, an economic overtone.

And now a digression. The idea that future stock dividends need be discounted represents the primacy in economic thought of *time* – this in contrast to physics, where the fundamental laws work no matter which way time is running. The objects of ordinary experience are divided into those that are time-bound and those that are timeless. A diamond is forever if only because nothing happens to it, however much it may look electrifying on the moving molded throat of the woman who is wearing it. Corporations, by way of contrast, are time-bound; they have a queer biological dimension and pass through recognizable stages in their existence. General Motors is often described as a *mature* corporation, as if it had a sleek forehead, silver hair, and a mistress at the *Pierre*; Apple Computer yet retains an adolescent air of sexual enthusiasm and appears prepared at any time hopefully to snort talcum powder, chug beer, or invite those girls in 104 over for a party. The law, it is true, absently invests every corporation with unlimited life, but of a sample of, say, two hundred American corporations flourishing in 1900, almost all have vanished, with only their names an object even of curiosity: *The All American Cracker Corporation, The Smithstown Foundry, Mother Hubble's Lard, Lydia Pinkham's*

Tonics. It is odd that so sophisticated and abstract an object as a corporation should be described by means of the ancient metaphors of growth, development and decay.

That digression has now been concluded.

Most stock brokers believe that a stock's price and its value are two different things. Set against this (or their) view is the thesis that stocks are worth what they are worth. Oskar Morgenstern and C. W. Granger, for example, write indignantly that "economic theory has been plagued for centuries with the effort to define the elusive notion of intrinsic value only to discover finally that there is no such thing." These are sobering words for the stock-market analyst (or for feminists fighting for various principles of comparable worth). What a stock is worth today is what someone will pay for it; what a woman is worth in the job market is what she can *get*. Morgenstern and Granger argue that there is an end to the matter. The intrinsic value of a stock (or a woman) is thus revealed as a fiction. *Hear, hear*, I am temped to say somewhat uneasily, but with perfect conviction. Still, something more than a stock's current price remains when its intrinsic value is discounted to zero. There is its *future* value – defined now as the price at which it *will* be selling. This way of looking at things does nothing, of course, to revive the idea that stocks, like blondes, have an intrinsic value that goes beyond their selling price; but it does place stock-market prices among the broad category of numerical objects that change over time, and that thus have a future and a past.

The Boys in the Back Room

Upon my promotion by means of marriage from loser to lover, I came to know on rather more intimate terms the community of brokers, from the boys in the back room performing lugubrious regressions on the wayward price of copper, to the ebullient and jolly fundamentalists in the window offices, covering each and every one of their bets with the hedge of their good-natured optimism. Given my own weakness for the crackpot in every category, I found myself irresistably attracted to *technical analysis* – the doctrine that the price of stocks (and nothing else) determines the price of stocks. The technicians that I knew were, for some reason, very mild and inoffensive characters, with watery eyes and unsuccessful moustaches, and an Ancient Mariner-like capacity to bore a visitor to death. Enter their office and ask innocently "What's cookin'?" and you were lost forever. Hearing that first fatal intimation of curiosity, a great wet smile of warmth and wonder creasing his features, Burt or Kurt or Rob or Bob would promptly rise from his chair and immediately – his behavior now an unstoppable blur of enthusiastic activity – fetch some immense chart from a rack, smooth the horrible thing on a table, and with brows furrowed commence an endless account in which the meaning of the figures would be explained in great and tedious detail. Some technicians shunned all forms of theory beyond the observation that the market moves in discernable trends. Others subscribed to a general form of almost medieval numerology and were very much taken by all sorts of numerical coincidences – the fact that when scrambled by some secret code the letters that make up John F. Kennedy's name express also the name of Abraham Lincoln, who, like Kennedy, was shot in the head by an assassin – and this is the really *weird* part – whose name is an anagram for Lee Harvey Oswald.

My technicians – I am thinking of them now as a clan – followed Joseph Granville's market letter and projected toward Granville himself a sense of admiration and gratitude so striking as to be indistinguishable from worship. This was not altogether surprising. From 1974 to 1981 Granville managed to call virtually every major stock-market turn; those brokers who followed his advice scrupulously made a good deal of money. So significant was his influence that for a time his own calls caused the market to turn, a superb and supreme example of a small tail wagging a large shaggy dog.

Just recently, Granville has put together an autobiography in which he expatiates on his theories and explains how the great things were accomplished. On a rainy Sunday not too long ago, I assembled a collection of his books, the autobiography included, and, after sinking into an easy chair (Almond Rocca to the left, cigarettes to the right), devoted myself to their study. I was surprised to discover a curious, boisterous, energetic, and truculent intelligence bouncing up from those books, a Hey-Look-at-Me! expression on its otherwise indistinct features. Although unprepared to cast his views in just this way, Granville (and all other technicians, I think; the affliction is generic) subscribes to a version of what orthodox economists call the *efficient market* hypothesis. Let us suppose (to adopt the Horatian mode), for the sake of illustration (and argument), that the stock market is regarded as a kind of computational device putting out the price of stocks on a daily basis. The external world impinges on this mechanism by means of discrete chunks of data: earnings reports, research results, boardroom gossip. Fundamentalists argue that this information is decisive and see the input into the market as a stream of slow moving, causally important, fundamental facts. The technician dissents only with respect to the speed of the stream. Information, he argues, travels at the economic equivalent to the speed of light and reaches every participant in the stock market at precisely the same time. Whatever is new in the news, the market has *already* reflected fully in the price of stocks. Adding fundamental facts to the historical record of stock market prices is adding by zero. Here, technical analysis has to some degree anticipated the direction of academic research, although the present point of similarity is, no doubt, an awkwardness for the academic, concerned as always with avoiding the friendship of his enemies. Yet if Granville, and other technicians, argued for the irrelevance of fundamental information to stock market prices, he failed to indicate *why* the market should discount such data, and thus left room for the academics to refine the thesis.

The efficient market hypothesis is pretty much a negative view of things. Public information is without value just because it is public. Granville's own theories are positive as well, and thus represent a healthful balance of Yin and Yang. In any market, he argues, a few traders are privy to large secrets. Such is *inside* information. If public information travels at the speed of light, inside information does not travel at all. The theory now acquires a certain lurid conspiratorial glow. Those who know the most buy when others sell or sell when others buy. In buying and selling, these operators leave a trail in the market, a kind of lingering fiscal scent. Like a beagle, the technician trained in such matters can investigate a historical record of stock-market prices, and by means of various *indicators,* pick up this smell. Knowing where the insiders are going, the technician proposes to follow at a discreet distance.

The details of Granville's theory are complex, although certainly not irrational; in many respects I have failed to do justice to the surprising energy and intricacy of his reasoning. As a whole, the theory is, of course, false in the simple sense that, when faced with all of the facts, it flops when the facts flip, and flips when they flop. Brokers who in 1982 sold on Granville's advice, sold in the face of a surging and magnificent bull market, and spent the years thereafter explaining to their clients how it was that they failed to make a boodle.

Mathematics, no less than economics, has always attracted a certain number of self-absorbed and clever cranks. Often, they are totally without formal schooling.

Abstracts of their passionate communications appear regularly at the rear of the learned journals. From time to time, they get something right. Professional mathematicians regard their efforts with sympathetic contempt. Academic economists, on the other hand, have *always* regarded the technician as a figure of fun. If he gets things right, the economist counts this as luck; if he gets things wrong, it is no wonder.

As for me? My own sympathies go toward the margins, where a variety of walnut-eyed characters and customers are lurking, fingers raised in admonition, their voices low, whispering of vast conspiracies, explaining how it is that things *really* work.

Physical, Material, Mechanical

Whatever it is that a stock-market analyst believes, what he must do (simply in order to make a living) is tell his clients whether stocks are on the way up or on the way down. This is a matter of forecasting the future. "Since prediction is one of the primary concerns of stock market analysts of all types," Morgenstern and Granger write, "it is desirable to state here [where else?] two fundamentally different methods of approaching this task." Their distinction is drawn in the context of a discussion of meteorology, but the point carries over nicely to economics. The first method, they continue, is this:

> Present to the computer a weather map of today, consisting of a set of numbers indicating air pressure, temperature, wind direction, speed, etc. Then instruct the computer to search its memory and to find the weather map most nearly identical to the given one. If today is, say, the 20th of August 1968, this may be the map of the 17th of July 1901. Then tell the computer to print out the weather map of the 18th of July 1901, which then will be used as the weather prediction for the 21st of August 1968. The second method is to neglect all past records of the

> weather, to determine the present weather map and use the numerical parameters in a mathematical model expressing our meteorological knowledge by a set of equations. Solve these equations numerically in order to obtain the prediction for tomorrow (in a computing time which must be significantly shorter than the time it takes for the weather to change).

Morgenstern and Granger dismiss that first method of theirs because "it requires no understanding of the mechanism which produces weather changes." This scruple, I observe (regarding the matter now as a debate), makes sense only if in knowing those mechanisms the weatherman is actually in a better position to tell the rest of us whether it will rain tomorrow. Despite the existence of very sophisticated hydrodynamic models of the atmosphere, the weather remains essentially unpredictable in those climatic regions where it changes rapidly (New York, say, but not Palm Springs). On the other hand, there are many physical systems that *cannot* be understood analytically in terms of an underlying mechanism – some so simple as a dripping faucet.

Still, Morgenstern and Granger are right to this extent: Science has been historically most successful when it dismisses the real world as an impertinence and treats instead a spare and elegant secondary system, something more or less cooked-up. Here the real world, if it appears at all, appears as a point mass instead of a planet. To the man in love with the texture of flesh, or those fabulous roiling clouds that Turner painted, or the smell of sweet and sour pork, the scientific world view must inevitably seem partial, dissective, unsatisfying; but what the scientist gives up with one hand (pleasure), he gets back with the other (power), so perhaps the scales balance after all.

The very simplest scientific description of a natural process involves nothing more striking than the observation that something happens when something happens

– the old in and out, consisting, in many cases, of two shaggy streams of data. Tapped on the patella, the patient responds with a twitch; impressed by a force, a ballerina accelerates; subjected to stress, a bridge, or a marriage, crumbles; given an input, the computer responds with an output – *Syntax Error*, most commonly. Between what goes in and what comes out is a device of some sort – hereinafter, *the* device, to abstract from the various possibilities. If all knowledge of its internal structure is blacked out, the device is known as a *black box*, a kind of enigmatic wedge standing like a rock in mid-stream between two temporal flows. With nothing else to go on, the scientist scrutinizing this setup attempts to draw a general connection, however vagrant, between what happens to the device and what the device then does. Such is the inductive impulse. Engineers and systems theorists refer to an analysis so limited as an *input-output* description. The psychologist concerned only with what the albino rat *does*, treats the rodent as a black box, and frames his theories entirely in terms of schedules of reinforcement – torture juxtaposed to behavior. In stock market forecasting, it is the technician, speaking strictly, who regards the market as an input-output device; in predicting future prices (one temporal stream), he looks only to past prices (the other temporal stream). With his nose to the charts, he remains indifferent to any patterns save the ones that he can see, and unimpressed by any causal powers beyond the powers of the pattern itself.

Two players at a chessboard define a *system* in a sense of that word too primitive to admit of further definition. The system consists of the chess pieces in a particular initial configuration, and the players themselves. For the most part, each player may be identified with the moves that he makes – a piece of reduction in which the elegant Capablanca remains as a series of linked and coordinated gambits long after the ephemeral Capablanca has departed for another match. One player moves; such is the input to the system. His opponent moves in response; such is the output. But the moves that a player *will* make cannot, in general, be predicted simply by a study of the moves that he *has* made. What more is needed? An account of the rules, of course, for these express the laws of the game, and a description somehow of the various *possible* configuration of the pieces. These configurations mathematicians refer to as a set of *states*. The set as a whole makes up a *space*. A *state-space* description is one in which the action of two temporal streams is mediated by a device in one of a number of distinct internal states. The mechanical models of Newtonian physics are imagined thus.

Now engineers (and the rest of us, too) think of machines in terms of heat, work, and energy. The panting steam engine is a paradigm. The living human mind finds life outside the body inhospitable, I would guess, but a computer may carry out its computations without ever being directly embodied in hardware. The Turing Machine is an example. And computers are typically state-space devices. In fact, state-space descriptions are evoked whenever ordinary differential equations come into play; they are thus part of the world's abstractions.

A machine, I think, can be defined as *any* device taking inputs to outputs by means of a finite set of internal states or configurations. But this idea is too general to be much use. For one thing, the definition simply specifies how a machine works, but not by what principles. This has always been fine by systems theorists, but not by me. Without content, there are no concepts. For another thing, the definition suggests that one and the same set of laws may apply indifferently to various machines. As it happens, machines, like biological species, have a riveting sense of their own individuality. It is only when the definition of a machine is specified that the doctrine of mechanics acquires explanatory force. And then, of course, one hardly needs the definition. I mention it only to establish a convenient point of reference.

In science, but not in life, an explanation is very often a matter of showing that what is, must be. The behavior of a machine is fixed entirely by what is done to it and its internal configuration. It is for this reason that a mechanical explanation is so very often deeply satisfying. To feel the beating human heart beneath one's fingertips is to touch a mystery. Much of the mystery (as well as the majesty) disappears when the heart is redescribed as a pump. The Newtonian vision of the world as a vast machine breaks down, as I have said, on the ragged right margins of our experience; but if mechanism is no longer a reasonable philosophy for the *whole* of Nature, it is a doctrine that retains its fragrant appeal elsewhere: in stock market forecasting, astrology, linguistics, artificial intelligence, catastrophe theory, meteorology, and molecular biology. Here machines figure as Markov processes, finite-state automata, Turing Machines, grammars, formal systems, ordinary differential equations, and finite difference equations. In each case, the simple intuitive idea of a device taking inputs to outputs by means of a set of states may be recovered – the gleaming skull beneath the glowing skin.

The triangulation of science by means of the concepts of a physical, material, and mechanical explanation is a rough and ready affair. Mathematics is neither a physical nor a mechanical science, and remains obdurately resistant to inclusion in the scientific world view. Mechanism has long since lost its mooring in Newtonian mechanics – a system like Methodism that is now intellectually closed. Machines need not be made of matter and exist apparently in that region of experience in which pure intelligence, and nothing else, dwells. Still, the coordinates are ball park, as cartographers say.

Bi-Coastal: West

At the western margins of the state of Washington, two great mountain ranges, running from north to south, stand to either side of an inland sea and a broad, alluvial plain. The snow-capped peaks of the Olympics lie between Puget Sound and the foggy Pacific ocean; some seventy miles to the east, past the cities of Seattle and Tacoma, the Cascades disappear into the mist, gray and crumbly.

The University of Puget Sound, which I had come to visit, stands on a splendid piece of property at the southern edge of the sound, overlooking a forlorn group of islands named after Indian tribes or animals – Vashon Island, Fox Island, Dog Island. I had been on campus for only a few hours, but I was already quite taken with the lowering sky and the languid sound and the somber Olympics on the far shore and the pebbly beaches and the murmurous, moldy forest with its strange mushrooms and the deep, creosote-woodsy smell of the place.

The chairman of the department of philosophy, John Magee, escorted me around, dutifully pointing out the sights – "That's our gymnasium, and that's Jones Hall, and that's our Hub . . . " – in a way that suggested shy pride and the worry that I might be looking down my nose on things in general. "I've been happy to spend my life here," he said wistfully, after fiddling with a pipe made of a metal column and a plastic bowl, which asbestos-adamantly refused to stay lit for more than a moment. I nodded manfully.

The University was composed of a series of cheerful red brick buildings with sloping roofs. Between the buildings exquisitely trimmed plots of grass put forth a napery of rain-washed green; elegant evergreens flapped and moaned in the breeze, a clutch of inky ravens taking flight every so often to wheel pointlessly in the sky and then land again on the topmost branches.

"Why do you want to leave New York?" Magee asked as we were walking toward Jones Hall.

"No Schvugies," I had said to a friend when asked precisely the same question. "You walk out in the street and there are no Schvugies walking around with one hundred pound radios glued to their ears."

For obvious reasons – Magee looked like the sort of man who might actually wish to clap some smoldering Black on the shoulder and exclaim moistly that he would be proud to think of LeRoy here as a friend – I elected not to pursue the theme of racial rawness.

"I've always been an educator, Professor Magee," I began instead. "I need a different kind of contact with my students, something more one-on-one, more intense, a little more intimate." Like a gentle westerly breeze, I let slip the implication that my own heart beat with a simple message: Get to know better those monstrously broad-shouldered students who stood posed glumly on the cover of the *Trailblazers* that Magee had sent me just a little while before.

Placing a very deft hand at my elbow, the gesture at once charming and old-fashioned, Magee propelled me the rest of the way across the quadrangle and up the steps of Jones Hall.

Inside, there was an enormous oil painting in which five robed young men stood solemnly in a group, their hands resting on a single, leather-bound Bible, which evidently functioned as a communicating liver or a single kidney. Magee identified the mummified quintuplet as our first graduates.

After that there was a portrait of some somber, very sullen, smoke-eyed figure dressed in black – an immense dignitary in the university's history – who, Magee informed me, had perished in a sailing accident. I imagined to myself this corpulent character sinking furiously beneath the salt-soapy waves of Puget Sound and like a tuna coming to rest on his parabolic stomach, all orders to make for the surface swallowed by the gloom.

Having entered Jones Hall from the west, we traveled peristaltically through a series of corridors until we arrived at the President's office in the east, before which a covey of twittering secretaries sat, so many sharp-clawed little birds.

Magee took his leave with a wan smile.

The President of the University bounded from his office in a wolfish lope and, after introducing himself, began pumping my hand with one of those athletic handshakes that in their action resemble an unstoppable piston. He looked searchingly into my eyes. He said he was very glad to see me, and did I have a good trip, and could I put out that cigarette? He seemed very fit, if somewhat nervous. I noticed that his smile and bow tie seemed to stretch in the same direction, giving him the unnerving effect of having two mouths.

"Come in, come in, come in," he said with a trochaic lilt.

From my perch on a green leather couch in his office, I could see a hedge outside, a worn but elegant sandstone railing, and another of those succulent late spring lawns. On the walls, there were any number of pictures of the President himself, beaming always, as he accepted foundation checks from pin-stripe-suited officials, or presented athletic awards to a gorilla troop of muscular young men, or posed genially with a trio of local boosters (his arms over their shoulders, like a penguin), who were obviously uncomfortable (tight little smiles) with the high company they were forced to keep. In one picture that especially caught my eye, the President appeared in his running outfit (shorts, a T-shirt, and very large, sparkling white shoes) as he circled around the academic quadrangle at an acute sideways angle to the ground, an expression of great effort on his perspiring face.

We talked of this and that. Mostly this. Sometimes that. Finally the pitch came. The President leaned back in his very large, burgundy office chair. The torch of knowledge, he gave me to understand, his hands touching at the fingertips to form an attenuated pyramid, having been lit by Socrates and Aristotle, and then passed to Saint Thomas Aquinas and Isaac Newton, had evidently been handed personally to the staff and faculty at the University of Puget Sound (by Lavoisier, perhaps, or Max Planck), and now stood flickering fitfully amidst the dripping evergreens. My task, should I accept it, would be to blow on it gently.

I said that I would be happy to do my best; but when I showed up for office hours just six weeks later – and here is the point of this exercise – there, outside my door, waiting apparently for me and no one else, was a haunting, luminous young woman, with high sculpted cheekbones, wide-set, chilly, grave, staring eyes, and small seashell ears, wearing dungarees and a silky black leotard stretched tight across her bosom, face and figure miraculously coordinated to express longing and a doomed, pathetic sense of loss.

"I want to study the philosophy of science," she said.

The Odds

On the one occasion in my life when I needed to make a statistical argument, I made an ignominious botch of the whole thing. I cannot remember the details, but evidently I confused one statistical test with another. As I endeavored to extricate myself from my predicament (chiefly by insisting that what I had said I had not meant, and vice versa), I found myself getting into deeper and deeper trouble (these things are like an especially insolent nightmare), until finally one bullfrogish biologist with belligerently trembling jowls rose slowly to his feet and, after waving a hand upon which a number of brown age spots had been inscribed, leaned forward and in a voice of doom said that I had no idea what I was talking about.

For reasons that are psychological, and that perhaps involve a touch of snobbery, mathematicians came late to a rigorous description of the theory of probability. It was not until 1931 that A. N. Kolmogorov provided a set of axioms for the subject. By then, of course, it was plain that statistical arguments played a crucial role in thermodynamics and quantum mechanics. Kolmogorov's work had the effect of bringing order to a subject that had been largely the creation of shifty but appealing gamblers at the French courts, and that had long exhibited, in consequence, an air of frank and preposterous improvisation. Mathematicians now regard the foundations of the theory of probability with a certain confident satisfaction; from an abstract point of view, the entire subject forms an aspect of the theory of measure.

Still, even though mathematicians have long accepted Kolmogorov's description of things, it has remained possible to ask, entirely on a level that precedes sophisticated theory, what it all means. The immemorial aim of the theory of probability, for example, is to associate numerical measures to stochastic experiments. If a fair coin is flipped, the odds of it landing on one face stand at 1 in 2. But a fair coin is *defined* as one which when flipped lands equally many times on either face. Probability has been expressed in terms of fairness; fairness, in terms of probability. It is invigorating to the philosopher to see perplexities emerge so early in the analysis of an idea.

Every science – every act of the intellect – rests on concepts that are assumed without explanation. In the theory of probability, these are the *elementary events,* which are intuitively depicted as the outcome of experiments involving roulette wheels, card deals, coin flips, and the like. In the case of a six-sided die, the obvious experiment has six possible outcomes. The numbers from 1 to 6 are points in the experiment's *sample space*. By long tradition, elementary events are assigned probabilities lying in the interval between 0 and 1. Zero corresponds to the event that cannot occur (the impossible event); 1, to the event that must occur (the ineluctable event). Each elementary event in a discrete sample space is assigned a probability – it is not the business of the theory to describe how – and by assumption these sum to unity.

The elementary events that result from throwing a fair die with six faces make for a *uniform* sample space in the sense that each outcome has precisely the same probability – 1 over 6. The points that comprise this sample space may themselves be grouped to form compound events. The six possible faces of the die, for example, constitute a *set* of elementary events; probabilities may be assigned to subsets of this set. The set as a whole is a subset of itself and receives the probability 1. The die must land on one of its faces. The subset composed of 1, 2, and 3 represents the possibility that the face value of the die is less than four. Its probability is 1 over 2. The empty set corresponds to the impossible event – the die does not land on any face and hangs petulantly in mid-air. Its probability is 0.

In the case of a coin, the experiment involves flipping the thing. The coin must land either on its head or its tail, but not both. The relevant sample space consists of two points: H and T. Flip a fair coin twice, and a slightly more complex sample space emerges: HH, HT, TH, and TT. The notation explains itself.

In throwing a fair coin twice, I generate two elementary events. Now the odds in favor of seeing the coin come up heads on my first throw are plainly 1 in 2. What of the second throw? Or the tenth, to imagine those throws continuing? They never change, those odds, and remain at 1 in 2, even if that coin has landed on one face for ten, or ten thousand, trials. The events are *independent*.

This statistical commonplace sets intuition to quivering. Odds that remain indifferent to the history of an experiment suggest that nature is carelessly without memory. Just so. To believe otherwise, as hunch-ridden gamblers do, is to commit the fallacy of the maturation of chances. Wherever the statistician goes in this argument, I am with him in spirit. I demur only with regard to the way in which he expresses himself. The statistician's invocation of a fallacy suggests that the gambler has somehow succumbed to a mathematical or logical mistake, a vulgar error, in any case. But the fallacy of which he speaks is not so much a mistake in reasoning (and hence not a fallacy at all) as a kind of bluff indifference to the statistical *facts*. What perusades the both of us (the statistician and me) to keep those coin-flipping odds the same is nothing more than experience.

On the other hand, if cards are being withdrawn from a standard deck, each deal *does* alter the subsequent probabilities. Here Nature acquires that wayward memory it so recently lost. On the first pass through a deck, the odds in favor of my pulling out an ace stand at 4 in 52. After freezing one ace from play, I have a 3 in 51 chance of selecting yet another. The probabilities are now *conditional*. If two fair dice are being rolled, to make the example more complex, the odds that their faces sum to seven stand at 6 in 36. Suppose, however, that the face value on either die must be less than five. What now are the odds of reaching seven as the sum of both faces? (It is astonishing how very quickly even elementary problems in probability achieve the complexity of a hairball.) Each die can fall in only four ways if neither die shows more than four – this makes sixteen ways in all. No other possibilities count. The sample space is thus truncated from thirty-six (the original sample space for all throws) to sixteen. Now the conditional probability of one event given the occurence of another is usually defined as a ratio – the number of ways two events might jointly occur over the number of points in the sample space. There are only two ways in which the face values of the dice could sum to seven if neither face shows more than four. Either the first die shows four, and the second, three, or the reverse. The right ratio is 2 over 16.

In flipping coins, the statistician confronts a sequence without memory. The probability associated with an event is fixed from the first and remains the same, regardless of what happens. This is an extreme case of conditional probability, in which no influence at all passes from one elementary event A to another elementary event B. The conditional probability of A, given B, is just the probability of A. The conditional probability of B, given A, is just the probability of B. The probability of their joint occurrence is the *product* of their individual probabilities. The odds that a fair coin when flipped will land twice on one face is 1 in 4. In the case of independent events, the formula for conditional probabilities thus collapses into a simple multiplicative rule. Nothing in the notion of independence, it is worth noting, requires that the elementary events be assigned the *same* initial probabilities. A coin may be weighted so that the odds in favor of heads, say, are 1 in 4 rather than 1 in 2. When flipped, this coin yet generates a sample space (although not a uniform sample space) of independent events.

Imagine a fair coin being flipped in succession, so that a longer and longer series of events emerges: HHT; HHTH; HHTHT . . . , for example. A sequence of this sort is known as a *Bernoulli sequence of independent trials*, with *identically distributed probabilities*. The Bernoulli sequences make for a first primitive model of a random process. The successes in a Bernoulli sequence are a record of the number of heads or tails in an arbitrary run. In looking at a large number of trials, the statistician expects to see that roughly half the outcomes are successes. For once, his intuition and my own coincide. The measured ratio of heads to tails, keeping now to the case of the fair coin, is a matter of the relative frequency with which one face of the coin appears in, say, ten thousand trials – 4526 to 10,000, to invent a plausible figure. Now this number is plainly connected to the probability initially assigned to the experiment – 1 over 2. The *law of large numbers* – there are actually several theorems: the lore of large numbers – establishes that, over the long run, relative frequencies and probabilities tend to coincide. This draws a satisfying and intuitive connection between random sequences and random experiments.

•••

I have described the elementary aspects of the theory of probability from the perspective of the mathematician concerned with setting out the conceptual details. The resulting view encompasses probability from the top down. To the statistician falls the unenviable task of dealing with probabilities from the bottom up. Quite typically, he is given merely the numerical record of some process and asked to pronounce on its stochastic character. Are the events that it records independent? Obviously related? Is the relationship strong? Weak? Statistically significant? Or what? I remember with a shudder being confronted once with a computer tape listing all fetal deaths for the City of New York over a twelve-month period. "See if you spot something interesting," urged a project supervisor much interested in air pollution and premature mortality. The thing is harder than it sounds. Having no access to the mechanism by which those numbers are generated, the analyst (me, in this case) is forced to deal with the evidence itself, a prospect a little like reading a newspaper in another language – Tamil say – with only a Tamil dictionary to go on.

The outcome of an experiment is an event or series of events. Frequently, the events of interest to the statistician (MacSuppes, to give him a name and endow him with a character) have no discernable (or accessible) numerical structure. If a fair coin is flipped twice the sample space that results is composed of the following points: HH, HT, TH, and TT. But suppose now that MacSuppes is interested in the number of heads associated to each point. Nothing in the sample space is numerical at all. To get anywhere, MacSuppes must endow the sample space with the properties of the real numbers. This endowment he creates by means of a mapping: HH→2; HT→1; TH→1; TT→0, for example. The mapping itself is known as a *random variable*, a somewhat confusing description inasmuch as the mapping is neither random nor variable. With the mapping fixed, the abstract objects that make up this particular sample space acquire a temporary set of associated numerical magnitudes. This allows the methods of mathematical analysis to be brought into play.

In this case – MacSuppes' Case – each outcome of the experiment gets the probability of 1 over 4. (The analysis that makes this clear is hardly a triumph of mathematical reasoning.) By the *expected value* of a random variable, MacSuppes, and everyone else, means the result of multiplying the value of the random variable for each outcome of an experiment by its probability and then summing the result: $2 \times \frac{1}{4} + 1 \times \frac{1}{4} + 1 \times \frac{1}{4} + 0 \times \frac{1}{4} = 1$, to continue the case of the coin. The expected value of a random variable is thus nothing more than a kind of average. The gambler betting on a fair coin flipped twice may expect to win precisely one dollar if he gets a dollar for each time the coin lands on its head.

It often happens that the statistician needs to measure the variance or dispersion of a random variable – the degree to which the values around the expected value are scattered. This is an item most naturally represented as the difference between each value of the random variable and its expected value. These differences are then squared to provide the statistician with a set of positive numbers. The (total) *variance* of the variable is defined as the averaged sum of these individual variances. This definition may be extended to cover the case of two random variables. The result is a measure of *covariance*.

Investigating a numerical sequence, the statistician is generally concerned with determining the degree to which one random variable – the number of deaths by cancer at a given age – impinges on another – the number of cigarettes smoked, say. The statistician is ignorant of the stochastic structure by which the variables are generated. What he is after is their *correlation coefficient*. As the term suggests, this coefficient indicates in a rough and ready way the extent to which two variables stick together. Technically, the simplest of all correlation coefficients between two random variables is defined as the ratio of their covariance to the square root of the product of their variances. When the relevant statistical operations have been performed, the result is a number – always a good sign in mathematics – that measures the degree to which the distribution of two random variables lie on one and the same straight line. This makes the concept of correlation contingent upon just one geometric shape, and lends a somewhat disturbing mathematical relativism to the very concept of correlation. Random variables that are weakly correlated under one definition may be strongly correlated under another.

Is there a connection between independence, which is a concept drawn from the theory of probability, and serial correlation, which is a concept drawn from statistical theory? There is. The correlation coefficient between two events A and B is zero whenever A and B are independent. The converse is false. If the correlation coefficient between A and B is zero, A and B may yet fail to be independent. Textbooks provide examples in which A is actually a function of B, but students and sociologists still flub the point with depressing regularity.

Russian Roulette

It is human nature to wish to connect the dots. Three centuries of scientific culture have reinforced the impression that the entire universe is at some level lawful, with the beat of its regularities sounding to the scientific ear like a great thudding heart. Astrophysicists and astronomers are committed to the idea of cosmic lawfulness, and for good reason: It is their stock-in-trade. But astrology, like palmistry, also rests in the end on the conviction that between the parts of the observable world there is a discernable connection. This need not be a matter of cause and effect. The grave coal-black eyes of the Gypsy fortune-teller, looking up in astonishment and alarm from an unhappy palm, have only read a sign, a visible marker pointing toward the catastrophe to come, not causing it. The contrary idea that events may not hang together at all is deeply repugnant. In quantum mechanics, surely the most depressing of the sciences, this possibility emerges as an epistemological principle, with electrons and subatomic particles bouncing around the place for no good reason that I, or anyone else, can discern. *I dunno, man, the wave packet just collapses*. Things that happen for no good reason are doubly dismaying in being unpredictible and inexplicable. The contemporary description of the subatomic world makes contact with an older, deeper, darker vision, in which irregularity, chaos, catastrophe, formlessness, and randomness are the somber and general standard.

Imagine the sequence of steps traversed by a drunkard. Leaning against a lamppost to begin with – the imagery here is as formalized as in the Kabuki – he lurches first in one direction, then in another. Aside from the fact that he keeps going, there is not much order here, or regularity, and no way to predict where the drunkard is bound on the basis of where he has been. Such is a *random walk*, a mathematical object with the unnerving capacity to suggest the whole of a life – my own, for example.

Looking at a drop of water beneath his microscope, the Scottish botanist James Brown noticed that small particles moved across the surface of his slide in aimless but agitated motion. The date is 1827. Brown conjectured that the particles were influenced by thermal currents. Were thermal currents at work, adjacent particles should have moved pretty much in the same way, like wood chips on a wave; in fact, particles separated by less than the diameter of a single molecule, we now know, move separately, in a nice display of waywardness and independence. Sixty years later, the French mathematician and physicist Jacques Perrin thought to analyse Brownian motion by means of a series of time-section graphs. Each particle that he tracked behaved with a vigorous and imaginative sense of sheer chaos, moving now in short choppy bursts, or drifting along the surface as if it were in fact following the long, listing walk of a drunkard with a severely compromised sense of direction. This suggested that what seemed on the surface to be random motion ran very deep, like a strain of melancholia.

In 1900, Louis Bachelier, a student of mathematics merely, and one of those dim-glowing figures in the history of thought who have never quite gotten their due, advanced the remarkable thesis that short-term price fluctuations on the French stock exchange could be modeled by a random walk. When I speak of modeling price fluctuations by a random walk, I mean to evoke an image of the statistician's glove over the market's hand. In many respects, Bachelier's theory of stock market price fluctuations *anticipated* Einstein's theory of Brownian motion, a circumstance that only the authors of French textbooks appear to recall. A glaring and unfortunate mathematical error distracted the attention of his teachers and colleagues. His work fell into desuetude, but the idea lingered. Sometime in the late 1940s and early 1950s, the British economist M. G. Kendall thought to look at the evidence instead of the theory. He collected over 222 stock market price series and subjected the data to quite elementary statistical analysis. What he discovered was an utter absence of statistically significant pattern. Prices moved as Bachelier thought they might.

During the 1960s, evidence in favor of at least some form of the random-walk hypothesis began to accumulate in large chilly drifts. *Primo.* At the University of Chicago, Arnold Moore studied weekly stock market price changes and found the fluctuations to be statistically insignficant. *Secundo.* His colleague, Eugene Fama, investigated stock market price changes that took place over one day to two week intervals – this over a period of five years – and concluded much the same thing. And, finally, Morgenstern and Granger, using spectral analysis to chart stock market price movements, concluded in the tough-guy, cigar-in-mouth, don't-kid-me-Bud style that they affected, that short term fluctuations were without statistical sense.

Still, this sort of evidence, although significant, does not imply that the action of the stock market is *entirely* capricious. There appear to be some long-term and stable statistical properties to price movement. For one thing, stock market prices have historically moved upward. If prices are under the control of a random walk, it is a random walk with drift. To the extent that it is drifting, it is not really random. There is a general tendency for stock market prices to reverse between trades. Certain points of reversal are favored. After several price changes in the same direction, price movements tend to persist. It is only when prices are charted over an interval from a day to three months that things seem random – and then only in the relatively limited sense that trading rules fail to outperform an arbitrary buy-it-and-hold strategy (the forerunner to the read-it-and-weep strategy).

In an efficient market information is discounted instantaneously. The claim that stock markets in the real world are efficient makes for one theory. In a market setting prices by means of a random walk, transactions are statistically independent. The claim that stock market prices in the real world obey a random walk makes for another theory. The facts themselves suggest only that stock market prices in the short term are weakly correlated, or not correlated at all. This makes for two theories and one set of facts, an irritating dispersal of conceptual resources.

In 1961, Paul Samuelson demonstrated for the first time that a genuine conceptual connection held between the efficiency of a market and the statistical character of its prices. This tied together two out of three items, which is not bad. The goal of a rational investor, Samuelson assumed, is to guess at the mechanism by which stock market prices are set. Those guesses themselves influence the mechanism, and the mechanism reflects the guesses, an interesting example of something like reciprocating double vision. In guessing at stock market prices, Samuelson's rational investor attends only to the past record of stock market prices. These he knows with certainty; the information is fully reflected in the price of stocks. What of future prices? Over, say, a period of a year? The rational investor has only the odds to go on, what statisticians call a *probability distribution* – the smear of a random variable. Thus an investor might guess that a certain stock stands a 1 in 4 chance of doubling its present value over the next twelve months, or a 3 in 4 chance of losing half.

This describes the general method by which each investor predicts the future; he guesses at the odds. It remains to be settled how future prices are actually set. Now, in general, what a stock is worth is determined by the demand for the stock, with IBM selling for a higher price than Purina Pet Foods simply because more people want its shares. Thus the here and now, What of the there and then? Demand is yet what counts, but with a difference. In the case of future prices, it is *anticipated* demand – what investors will bid and not what they do bid – that causes the price of stocks to slow or surge. And the rational investor, Samuelson assumed, will tend to bid prices to their mean or expected value. Buying today, this optimistic character bids prices to the mean price he expects will prevail at the end of his investment horizon – twelve months, say.

All this makes for a very modest budget of assumptions. The sceptic, determined to see statistical order in stock market prices, might well find himself scrupling at little so far – a classic case of a man petting a Puma because he has failed to observe its teeth. Stock market prices, Samuelson was able to demonstrate, actually exhibit little or no serial correlation. On the average, a set of such price sequences behaves with no statistically significant movement in any direction.

This is a powerful and disturbing result, but not one that yet implies directly that stock market prices are under the control of a random walk. Samuelson's argument is statistical in its conclusions. To assert that stock market prices display little or no serial correlation *because* the market describes a random walk is to take an additional step toward a theory. But, then, what better theory is there?

This thesis (the very idea, in fact) filled (fills) most brokers with deep and understandable anxiety. Stock market technicians such as Joseph Granville reacted with florid indignation when confronted with the random-walk hypothesis. In this reaction, there was a nice irony. To the extent that Granville and others discounted the importance of fundamental facts, they adopted in advance one aspect of the efficient market hypothesis and made early contact with later academic research (a feat possible only in a universe in which time travel is allowed). Academics went further still and argued that, since the behavior of stock market prices was random, no facts whatsoever came to influence the formation of prices, especially not facts about past prices. This placed the technican in the unhappy position of a man who manages to see the ultimate consequences of his own argument reach out and bite him squarely on the ass.

A Farewell to Physics

Moshe was a mathematical physicist: It was a treat to watch him drink tea. He was short and immensely fat, with a great spherical head over which a few coarse, black hairs had been curled forward in a dainty Caesar cut. Sitting at the edge of his chair, his legs crossed at the ankles, he would incline forward in a wedge, lift the delicate cup and saucer from the table, a heavy gold identification bracelet slipping along his hairy wrist, and after ceremoniously stiffening his pinky, take a creased-lip sip that was a marvel of chaste delicacy. The operation complete, he would sway gently forward to replace the cup and saucer on the low, rectangular table, flow upward to the vertical, and then dab at his lips with a linen handkerchief that he had plucked from the tube of his jacket sleeve.

In some former life or lore, Moshe had been a tank commander with the Israeli Army; the tea ceremony he had learned in Paris. But the Karma by which he had escaped from the Middle East was incomplete or flawed or impossibly inefficient. At Dauphine, where he taught physics, his classes consisted of students with deep, hopeless, melancholic eyes who had come to Paris from the French-speaking third world in order to obtain a technical degree by any means possible, and who impudently regarded Moshe as nothing more than yet another uninteresting, tubby obstacle. For his part, Moshe acknowledged that he tended to evaluate his student's work (and their worth) entirely in terms of the thickness of their eyelids and the oiliness of their skin.

"These people," Moshe said, his hands splayed over his knees, "cannot even write a simple sentence in French."

He reached discreetly for one of the buttery cookies that he so far had managed manfully to resist.

"Do you know what they say to me when I return them their examinations?"

Moshe paused to wipe the crumbs from his lips.

"Tu es raciste," he squealed indignantly.

Je m'imagine cela.

Toward his colleagues in mathematical physics, Moshe brought the same resources of musical contempt. Lowering his voice to a rumbling bass, an expression of serene composure on his mobile face, and speaking both French and English as if they were dialects of some common primitive language to which he alone had access, he declined a series of sibilant adjectives by order of distaste: senile (his own thesis adviser), servile (his junior colleagues), silly (his critics, the world), stupid (his collaborators), syphilitic (Professor X). Famous figures in contemporary physics were revealed to have astonishing intellectual limitations, but contemporary physics itself Moshe treated reverently as a frantic, fantastic saga; his description of the various schemes for grand unification called to mind nothing so much as those syncretistic religious bazaars of the ancient world in which spokesmen for the Great Mother would advert to her prowess while, companionably nearby, others would advertise the virtues of Osiris, or the Sun God, or the dark Assyrian gods with rolled beards.

"No, no, no," said Moshe vigorously when I asked whether physics had disappeared as an intellectual way of life; but afterward, when we were on the street, I asked, "Do you really believe in all that stuff?"

"Of course not," he said promptly, shifting the bulk of his body gracefully from one foot to the other. "That's just the party line."

From Bad to Worse

Thermodynamics is one of those nineteenth-century disciplines that play largely a polluted role in the history of thought. All too often the concept of entropy, for example, pops up where it does not belong, like some idiotic dog stiffening in excitement at the least opportune moment – in economic theory, or policy analysis, or while leaning amorously against the booted calf of a glowering Hispanic deliveryman. Even physicists feel edgy and suspicious when the matter comes up for discussion, and generally wish that the whole subject would fade away, like the now mythical ether. When Ilya Prigogine won a Nobel Prize for his work in thermodynamics, the reaction among mathematicians was thankfulness that no one in a position to disburse funds, or anyone else for that matter, had any idea of what he was talking about.

The first law of thermodynamics states that energy may neither be created nor destroyed and functions as a kind of cosmic principle of accounting. What is gained here must be lost there. On the quantum mechanical level, the first law is apparently false, or at least not clearly true. Some mathematical physicists now believe that the universe may have emerged spontaneously from an unpredictible ripple in the primordial quantum field. Still, whatever the details (or the truth), the first law of thermodynamics is plainly something with which a man might live without too much wringing of his hands. Not so the second law. The *entropy* of a body (any body) nineteenth-century physicists defined as the ratio of an increase in its heat to its temperature. In a closed system – a coffin, say – the entropy of the system itself may never decrease, and, in fact, tends toward a local maximum. In this sense, the heat available for work is insistently undergoing degradation. The whole thing goes from bad to worse.

Not that anyone needs thermodynamics to reach this

conclusion: The second law arises in physics as a frank inference from experience. The Wagernian air that the law has acquired in popular consciousness comes about when the principle is generalized. The universe-as-a-whole is quite obviously the whole of the universe – what else could there be? – and thus constitutes a closed system. Here, too, the entropy of the system is tending steadily toward a depressing maximum, with the heat available for work (and thus work itself) insistently undergoing degradation. The inevitable result is apt to be something that Walther Nernst called the *heat death* of the universe, a state of equilibrium in which degradation is at a cosmic maximum. I have always thought of cosmic heat death in terms of those flaccid jacuzzis running in very bleak health clubs, and look forward to the end of time with a shudder.

Physicists originally couched the analysis of entropy in terms of heat and temperature. Until the advent of the atomic theory of matter, heat remained impalpable, a kind of sinister theoretical fluid leaking from one body to another. On the atomic theory, heat is defined as a function of the *motion* of atoms – their mean kinetic energy, in fact. This serves to tie thermodynamical to mechanical concepts. But the principles of thermodynamics lend to things an ominous direction in time. Things go from bad to worse, and they go in only one direction. Those Newtonian particles bounce around with no decent sense of where they are going or where they have been. The atomic theory of matter clarified the concept of heat; it did little to show how thermodynamic principles might occupy so prominent a place in a Newtonian universe.

•••

Sitting at my desk, at ease, I watch a great blue cloud of cigar smoke drift upward. In time, the smoke comes to occupy pretty much the *whole* of the room's limited space. This is just what I expect.

When the physicist speaks of smoke, he means a swarm of atoms; different swarms make for different *micro-states*. These the physicist distinguishes by means of the position of their atoms. Quite obviously, there is no getting at any particular micro-state experimentally – there is too much to see, too little time. Yet on the Newtonian scheme of things, the molecules that make up the smoke presumably move without any apparent plan or purpose. Where they go is a random matter pretty much of what they hit – each other or the walls or ceiling. There is no special statistical reason for the molecules to be in one place rather than another. Why, then, do they refuse so adamantly to collect in a corner or arrange themselves into some goofy shape in space? The incompatibility between what one actually sees and what one might expect is known as *Boltzmann's paradox*, an unhappy name, if only because no real paradox is forthcoming, but an unhappiness nonetheless.

The tight, tense little circle of argument that I have just recounted suggested to Ludwig Boltzmann, who wrote and worked at the end of the nineteenth century, the inadequacy of a purely micro-static analysis of physics. Common sense and our common experience, he reasoned, may be brought into alignment only when various micro-states to a physical system are *classified* by macroscopic parameters: temperature, pressure, volume, and the like. Two levels of analysis now make their fateful appearance in physics, just as in biology or economics.

Suppose, for example, that my room is divided in half by a partition. Precisely four atoms are imagined; with the atoms go three overall patterns of classification. Those atoms may find themselves standing together to the left or to the right of the partition, or arranged so that three atoms fall to one side of the partition and only one to the other, or pairs of atoms may be matched to pairs of atoms on either side of the partition, as in doubles at tennis. If the atoms are scattered *randomly* – by being dropped from a chute, say – *each* distribution is alike in terms of the probability of its occurence – one over sixteen, as it happens. Now common sense tells me, in its usual seductive, wet whisper, that what I am *un*likely to see in this set-up are four atoms aligned together as a quadruplet. And for once common sense is right. When weighed against *all* other distributions of the atoms (fourteen, in all), *that* distribution is statistically unlikely.

In this case, of course, common sense comes by its conclusions only through the intervention of its friends. The contrast achieved between one arrangement of the atoms and those that remain is made possible only by means of a classification imposed from the first; without the classification, there would be no sense to the contrast. I am the last person to scruple at any scheme that seems to work, and what works for four atoms works for all the rest. To the extent that the smoke from my cigar occupies in time the whole of the space available to it, the distribution of its atoms is *uniform* – this in the sense that each of the atoms in the smoke is at roughly an average distance from its neighbors. In imposing a classification here, the physicist sets the uniform states against all the rest. By means of an obvious counting argument, this very simple act of conceptual segregation brings into alignment his expectations and the facts. The uniform distribution of atoms is statistically favored: *It is thus most likely to occur.*

•••

The resolution of Boltzmann's paradox closes the connection between the concepts of thermodynamics – heat, work, entropy – and those concepts needed to analyse the imagined universe of statistical mechanics. Entropy, in particular, now admits of expression in probabilistic terms. To the extent that the entropy E of a system reaches a maximum at equilibrium, E should *in general* be proportional to the number of distinct micro-states realized by a given macro-state, and proportional thus to the probability of the state itself. Whatever the macroscopic state S of the system, Boltzmann reasoned,

$$E = k \log \text{probability } S,$$

where k is Boltzmann's constant. At equilibrium, the number of distinct micro-states is at a maximum. And so too is the entropy. The smoke from my cigar occupies the whole of the room just because so many more possible arrangements of its elements are compatible with the fact that the entropy of this little system is tending toward a local maximum.

It is unsettling (to everyone but the philosopher) that a gain in clarity in one place is sooner or later offset by an eruption of inkiness in another. This, too, is a principle of conservation. Physics has always been vulnerable to a certain red flicker of hot temptation – the urge to explain the actual in terms of the possible. Critics of Boltzmann's famous H theorem thus argued peevishly that Boltzmann succeeded in demonstrating only what he had first assumed. The laws of thermodynamics he deduced from statistical mechanics by means of a physically alien set of concepts – those drawn from the theory of probability. To this line the mathematician in me responds with an indolent So What? Still, the arrow of time that Boltzmann thought he finally read in statistical mechanics tends actually to move in great loopy spirals. The laws of thermodynamics, as I have said, are anisotropic. They go in one direction – downhill. They are thus qualitatively in conflict with the rest of Newtonian physics. Statistical mechanics provides a brilliant and persuasive explanation for thermodynamic laws; and yet Henri Poincaré was able to demonstrate, by means of an absurdly simple proof, that *any* statistical–mechanical configuration, of whatever degree of implausibility, is bound to *recur* at some far reach of time in all its vividness, poignant symmetry, and daunting complexity. Physicists often explain this result by observing only that the time involved in recurrence is very long. No doubt.

Like Georg Cantor, Boltzmann died deeply depressed over criticism of his work. To the end, he retained an untroubled faith in the law that he discovered. He even had the thing engraved on his tombstone. Once while living in Vienna I wandered past the cemetery where he lies buried, simply to have a closer look and pay my respects. It was spooky to see those symbols carved into the marble, especially in a city where things went from bad to worse more rapidly than even Boltzmann might have thought possible.

Information

The American mathematician Claude Shannon created the theory of information in 1948. His papers, which he prepared for the Bell Research Laboratory, were mathematically straightforward but conceptually daring and deep. For a number of years thereafter, social scientists entertained cordially the hope that in the theory of information they might have access to the discipline that allowed them finally to treat social life by means of concepts as rich as those found in physics. In this regard, they had been disappointed before – by cybernetics, game theory, statistical decision theory – and were destined to be disappointed again – by systems analysis, graph theory, matrix algebra, input-output analysis and theories of complexity and communication. Of information, of course, there is always too much, and too much of what there is is of little or no use. But the stuff that is so often touted by IBM is *interpreted*, facts and their factoids, something with *semantic* significance, in any case. Shannon was interested in information from a *syntactic* point of view. Small wonder, then, that when the social scientists mastered the mathematics, they proclaimed themselves disappointed with its importance.

To begin at the beginning: Since the time of Boltzmann in the nineteenth century, physicists had suspected that a connection of sorts held between the concepts of entropy and information. What precisely the connection was, they could not say.

The date is now abruptly a half century later and the middle of the muddle. The attempts to illuminate the concept of information by means of the physics of the matter, although they bulked large, weighed little. Nonetheless, strategists for the Bell Telephone Company, confronting a post-war world with a great animal appetite for communication, quite naturally wished to know how information – whatever the damn thing was – might best be communicated over a channel – a telephone wire, for example. By "best," they meant most efficiently; by a "channel," any device taking messages from one place to another.

The waves of greed and genius are now about to meet and merge.

A *communication system* Shannon defined as a conceptual object of four parts. First, there are the *messages* themselves: discrete bits of data in the form of letters, words, sentences, or numbers. Then there is the *source* putting out messages, the *receiver,* who gets them, and the *channel* over which the messages are sent. Now, in the very process of communication, something inevitably gets lost. There are electrical disturbances, stray noises, mysterious glitches. (Indeed, just recently, two physicists, fine-tuning their astronomical radios, concluded that the inevitable, ineliminable hum that they heard represented nothing less than the cosmic cackle of the Big Bang.) Shannon was interested in the stuff that got through despite the noise. And this he *defined* as information.

In the simplest of set-ups (and no more is required), a single message might consist of a single letter, a sequence of messages of a sequence of letters, separated, perhaps, by marks of punctuation. Whatever the source, each message that can occur occurs with a fixed and definite probability. What is being measured is thus the information *of* the message and not the information *in* the message. Now here is the lovely, delicate point. There is a perfectly obvious connection between information, uncertainty, and probability. The information conveyed by a message (or messages) varies inversely with its probability, and serves as a measure of the uncertainty relieved by communication. Having gotten an unlikely message, I have reduced a large area of intellectual uncertainty to zero. Simple, no? And elegant?

But this is to present the picture in the large; the delight is in the details, where a *specific* measure of information may be found. The logarithm of a number is just the number required to represent that number as a power of ten (or some other number). The logarithm of 100 is 2: ten to the second power is 100. Ten to the first power is 10; and 10^0 is 1 even though the idea of multiplying ten by itself zero times is rather odd. Numbers between 0 and 1 are represented by exponents that are at once negative and fractional. In this interval, their logarithm must obviously *decrease*. Probabilities take values between 0 and 1. This suggested to Shannon that the information I assigned to a message X might simply be a function of the logarithm of its probability:

$$I = -\log \text{probability } X$$

The negative signs cancel. The measure is positive. As the probability of a message approaches 1, the information that it conveys approaches 0. The measure is natural.

The formula itself is much like the one Boltzmann used to describe entropy. The whole thing is spooky.

I have explained the concept of information in terms of logarithms taken to the base 10. The base 2 is as natural, and more convenient. In this system, all numbers are expressed in binary form as a sequence of 0's and 1's. In order to specify a single symbol in a binary alphabet, exactly one symbol is needed: either 0 or 1. It is useful to think of 0 and 1 as making up *bits* – the smallest unit of information. The number of bits in a string of symbols is then just the number of its symbols. From another point of view, this terminology, and the concept that it suggests, has a deeper justification. Consider a message whose probability of occurrence is precisely .5. Its logarithm to the base 2 is -1, its information content, 1 bit. *The* unit of information thus emerges as the information presented by the appearance of one of two equally probable messages.

Given the concept of the information of a message, it is easy and natural to define the more general notion of the information resident in a set of messages – an *ensemble*, as mathematicians say. This is just the information of the various separate messages weighed against the probability of their occurrence. A (general) information measure is *additive* if the information associated with two (or more) messages is the sum of the information in each. For the additive measures, it is possible to *prove*, and not merely assume, that Shannon's measure of information is unique: No other measure will do. This circumstance lends to Shannon's definition of information a certain enviable authority.

Information, on Shannon's view, receives an adequate mathematical representation only by means of coordinating concepts drawn from the calculus of probability. In ordinary language, the information of a message is a function somehow of its *content*, and content is a semantic concept, one that receives no amplification from Shannon's theory. I mention this only to enforce a sense of how strange it all is. From the definition of information, it follows almost at once that the information of a statistical source is greatest when each of its messages is equally likely. It is this fact that suggests a queer point of contact between information and entropy – two concepts with what on their face appear to be a frosty dissimilarity. The entropy of a physical system, Boltzmann observed, is at a maximum at precisely that macroscopic state compatible with the greatest number of distinct micro-states; in those statistical sources generating the greatest number of alternative messages, information reaches it own rosy limit. To the (very partial) extent that it makes sense to talk of information and entropy together, an increase in the entropy of a physical system corresponds to an increase in its information, but a decrease in its *accessible* information, corresponds to an increase in uncertainty.

•••

We all of us live in the perishable Now, the future and the past abstractions alike. Only the human intellect, and not any of its senses, is given to roaming the corridors of time, searching backward and forward for pattern and order and a busy sense of purpose. How disappointing in this regard to run up against randomness, which, when it is applied to stock markets or dice or life itself, makes for no better explanation of things than That's just the way it is, kiddo.

In looking at a roulette wheel, the statistician sees a sequence largely without pattern; the gambler see a pattern, or thinks he does, largely without purpose. No wonder these Las Vegas types so often take their own lives. Einstein could never reconcile himself to the idea that electrons jump in what amounts to a spirited display of free will – what quantum mechanics suggests – and to the end of his life regarded quantum theory as an abomination. This is a conclusion to which a man might give his assent without sharing any of Einstein's classical scruples.

Still, the relationship between randomness and those large metaphysical items, cause and determinism, is subtle, surprising. In applying probabilities to poker, the gambler gives up the all-too-human hope of tracing specific causes to specific effects. Random processes are not deterministic – at least not on ordinary levels of analysis. This might suggest (by contraposition) that deterministic processes are not random. Consider, then, the decimal expansion of *pi*. A finite estimate of *pi* may be contrived by dividing the circumference of a circle by its diameter. But *pi* is itself irrational and cannot be expressed completely in this way: The decimal expansion of *pi* goes on forever. Wishing to know how the decimal looks, the mathematician makes use of a variety of algorithms. These serve to compute *pi* to any desired degree. The operation of an algorithm suggests a deterministic process if anything does. Yet the sequence of numbers expressing the decimal expansion of *pi* obeys every ordinary statistical test for randomness and appears on the printed page to be utterly without sense.

And here my argument acquires a second, glittery layer of complexity. From the point of view of the mathematical theory of information, information and redundancy are dual. The more information in a message, the fewer the repetitions; the more repetition, the less information. The analogy to thermodynamics is pointed and exact, if limited and partial. Now consider a binary sequence of numbers that is random on all statistical counts. In a sequence of this sort, the moving eye hunts for pattern but finds none – a case of pure formlessness, or so one might think. Not so. The theory of information implies that is it precisely a sequence of this sort – and no other – that contains the very most information.

The ordinary magician manages on occasion to prompt the ace of spades to make an unexpected appearance at the top of a deck of cards or prompts a pallid pigeon to peep from the interior of an empty top hat. The mathematician, who belongs on precisely the same stage, follows with a greater, grander trick: By means of a mathematical sleight of hand, *he* erases the line between what is random and what is rational.

That trick involves no paradox, of course. It is a puzzle nonetheless.

My Sign

Most of the middle-aged men I know, and almost all of the women, regard dining as an exercise in self-denial. On coming into a restaurant, they look at the dessert tray, their eyes glistening, or at the french fries and sirloin on someone else's plate; but when it comes time to order, the men think of cholesterol or coronaries, and the women pat their hips and remember just how they looked in that bathing suit (the one that looked *so* good in the photograph in *Vogue*), and murmur to themselves that they'll simply die if they put on another ounce.

"I'll have the salad, please."

"Just the salad?" – this from chirpy Hi-my-name-is-Sandy-I'll-be-your-waitperson.

"Just the salad, Sandy."

"La salade," or *"Nur ein Salat,"* to cover the abstemious calls as I have heard them in Paris or Vienna (where, of course, the order is taken by a man).

"And a Perrier."

It was for this reason always a special pleasure for me to watch my friend Leo Rubble inspect a menu. The listed items he plainly regarded as representing an embarrassment of riches. With his head cocked at an oblique angle, he would undergo a marvelously visible struggle in which the various dishes would be vicariously tasted and then compared, the whole process rather resembling the efforts a man might make in determining a selection at a bordello. He would then order, with a careful accent to his execrable German, a sequence of at least four courses, one that culminated in a dessert of epic stature (a huge, gooey pudding, or something involving a great deal of whipped cream).

For all that, he was thin. Waiters approved of his bill of fare; women urged him to finish every bite.

We were sitting together in a restaurant in Vienna that was well known for being well known. The walls were pale yellow, the tables and chairs arrayed side by side along a central corridor.

Leo ate with adolescent industry. He was dressed casually, but in very expensive clothes: well-cut slacks, a knit shirt open at the collar, Italian loafers. His face was neither handsome nor striking, but infinitely soothing. His dark coloring and heavy brown hooded eyes with their thickened lids gave him a look of sensuous melancholia. With my own salad lying in a soggy heap, I twirled a toothpick, and lit a cigarette, and played with my fingers, and looked out at the pretty woman who were daintily eating white asparagus, or running their tart tongues over their voluptuous lower lips, or doubling their chins to examine an imaginary hair or piece of lint on their generous bosoms, or simply smiling discreetly in an arch, superior, satisfied sort of way, while their swarthy escorts gabbled to themselves in Italian or shouted *Ciao* from one table to another.

I ordered an espresso. From the bar at the head of the restaurant came the peculiar sound of laughter in another language.

It was impossible, really, not to experience a sense

of well-being in Leo Rubble's company. He was a man on the friendliest terms with fate. Like an eel slithering through seaweed, he had a kind of genius for moving acquisitively through life. Somehow things came to him. He lived in a wonderful flat in the first district, for which he paid little rent, his services to the landlord's wife excepted. He owned not one but two Lamborghinis, low, flat, sinister Italian sports cars said to be capable of exceeding the speed of sound, and a BMW, which he had acquired as the result of a transaction involving everything but money. Austrian university officials pressed him to lecture their charges, and regarded his inability to speak German as nothing more than a majestic whim. His girlfriend, a gentle, trusting, lovely, limpid creature, with straight hair, black eyes, a slightly snub nose, and full lips, adored him plainly, and would have bowed when he entered the room, had she thought it would give him pleasure.

"Well," he said, looking out vaguely at the smoky room, a glaze of contentment over his eyes, "it's my sign."

Later that evening Leo took a very large volume on astrology from his bookshelf – we had returned to his flat – and read aloud a description of my own sign: moody, impulsive, romantic, brilliant, poetic, mathematical, aloof, distant, rebellious, iconoclastic, generous, and kind.

Not bad, I thought. What about the rest?

Leo smiled enigmatically.

"A disaster," he said.

Kolmogorov Complexity

The aim of science, René Thom once remarked as we sipped espresso in a café near the Opéra, is to reduce the arbitrariness of description. I nodded my head and fixed my face in an expression that I thought conveyed a sense of alert but sophisticated appreciation. On the view of things that I had been taught at Princeton, the aim of science, insofar as science has any aims whatsoever, is a matter either of explanation or prediction. Going further, explanation involves seeing in the particular (this swan is white) intimations of the general (all swans are white) in such a way that the particular, when properly described, follows deductively from the general. In moving upward (past the swans, at any rate), the scientist ascends toward the laws of nature.

I thought to ask Thom what he meant, but by the time I had posed my question in a way that suggested I knew the answer, Thom was industriously applying himself to an apple tart.

Science is pursued, I think, for many reasons, not the least of which is to fill up the time. In this regard, it is always successful. Insofar as science is purely an abstract activity, like mathematics, chess, or nuclear strategy, it is undertaken chiefly for the acquisition of that magical moment in which things that formerly stood distinct and separate fall together in a limpid whole. Such is intellectual bliss – paler by far than physical bliss, but nothing to sneeze at either.

Blisswise, certain concepts form a tight circle in the sense that each may be used to define and justify the other. *Complexity, information, randomness, order,* and *pattern* (or *form*) are connected like the members of a family of cheeses: Gruyère, Brie, Port Salut, Camembert, but not Velveeta. This suggests that they may all flourish, or fall, together. Shady characters of all sorts – semioticians, anthropologists, linguists, sociologists, communication theorists – are especially partial to concepts of just this kind, perhaps because of the way that their names fill the mouth when uttered. This is no reason to reject these concepts out of hand, but no cause for congratulation either.

The technician, or the astrologer, no less than the rest of us, is pattern intoxicated. Reading the charts or the stars, he sees the subtle seams by which nature is constructed – the pattern at the bottom (or top) of things. A pattern is peculiar in that knowing part (moving in) one is likely to guess correctly at the rest (moving out). The patterns scouted by the stock market technician are especially plain: If they are there at all, they are there on the surface of things. On the other hand, consider the numbers 1, 4, 1, 5, 9, 2, 6, 5, 3, ... There is not much by way of pattern here; still less when the sequence is extended: 5, 8, 9, 7, 9, 3, ... A cursory examination might suggest that these numbers are quite without significance. Not so. They represent the decimal expansion of *pi,* to use an example that, like Mexican food, keeps coming up. Here the pattern is a matter of the way in which the sequence is *generated,* and lies hidden from the surface.

There is pattern, then, and generative pattern. Suppose the world contracted to a pair of symbols: 0 and 1, say. A binary sequence is a system of such symbols in a distinct order – 0, 1, 1, 0, 1, 1, for example – and of a specified length – six in the present case. Six binary symbols may be arranged in 64 separate sequences. In the general case, a sequence of length n (there are n symbols) may be recast in 2^n separate ways as 2^n separate sequences. Sixty-four is just 2^6, where 2^6 is 2 multiplied by itself 6 times.

Imagine now that binary sequences are being produced at random – by the action of a roulette wheel, for example. Of the two sequences

1) 0, 0, 0, 0, 0, 0,
2) 1, 0, 0, 0, 1, 1,

the first seems distinctly less likely than the second: A man idly flipping coins does not expect to come up with a run of six heads. Yet in point of probability, the two sequences are reckoned alike. There are 64 possible sequences in all. Each has a 1-in-64 chance of occurring. The most natural probability distribution over the space of n-place binary strings assigns to each string the same probability – 2^{-n}. It goes against the grain, mine, at any rate, to accept this conclusion, especially when n is large; but nothing *in the sequences themselves* indicates obviously that one is less (or more) likely to occur than any other.

Sometime in the 1960s, Kolmogorov (the same Kolmogorov, by the way) thought to argue that the degree to which a given binary string is random might be measured by the answer to a simple question: To what extent can the string be re-described? Kolmogorov thought of the possible re-descriptions of a given string as instructions to a fixed computer. Now if S is a binary string its length is measured in bits. An n-place binary string is n bits long. The most obvious re-description of S is S itself – the sort of thing I might send you to make sure that you get what I mean. In the case of sequences such as 2, nothing less will do. 1, on the other hand, may be expressed by a single terse command: Print 0 six times. A simpler description of binary string is thus a *shorter* description of the string. Sequences that *cannot* be generated by shorter sequences, Kolmogorov argued, are *complex* or *random*. This is a definition. But random sequences are precisely those that are rich in information. The definition thus ties together four concepts loitering casually at the margins of this discussion: randomness, compactness, complexity, and information. Playing an unusually inconspicuous role is the notion of probability.

Kolmogorov first spoke on this subject in a brief note published in 1967. His work was duplicated by the American mathematician Gregory Chaitin, who experienced a flash of intellectual lightning while an undergraduate at the City University of New York, sitting among students baffled by long division. The subject is known now as *algorithmic information theory.* Those algorithms are a reminder that Kolmogorov thought of descriptions in terms of inputs to a fixed computer.

Quite surprisingly, the problem of decisively determining whether a given string is random turns out to be unsolvable. If a shorter description of the string may actually be produced, well and good. If not, all bets are off. A shorter description may exist; then again, it may not. There is no demonstrative telling. The *decision problem* for complexity is recursively unsolvable. Like truth, randomness is a property that remains ineluctably resistant to recursive specification.

Kolmogorov's elegant and simple idea – a little jewel, a diamonoid – achieves its startling effects by means of an especially simple series of inferences. If all else fails, a binary sequence of length n may be re-described by a binary sequence of just the same length. There are 2^n such sequences, and $2^n - 2$ sequences shorter than this. But on any reasonable interpretation of complexity, sequences within a fixed integer k of n itself must be reckoned random or complex if the n - place sequences are themselves reckoned random or complex. It follows that only $2^{n-k} - 2$ sequences are less complex than $n - k$. If $k = 10$, the ratio of 2^{n-10} to 2^n is precisely 1 in 1024; the ratio of simple to complex sequences is thus on the order of 1 in 1000. This means that of 1000 sequences of length n, only *one* can be compressed into a program more than 10 bits shorter than itself. The number of purely random sequences grows exponentially with n, of course, and this implies that randomness and complexity are the norm in the general scheme of things. But if *most* sequences are random, the appearance of 1 *should* prompt a natural sense of surprise; sequences like 2 are what one expects and what one generally gets.

This line of argument, of course, resolves one problem only by embedding it within another – resolution by delayed dismay.

Metaphysics, the reader may have guessed, is not entirely in my line. I am tempted to cross over just this once. On Kolmogorov's definition of complexity, sequences are simple if they can be briefly described. Science is itself a matter of data; success in science a mystery of abbreviation. A law of nature is (among other things) data made compact: $F = ma$, said once and for all, the whole of an observed or observable world expressed in just four symbols. The fact that science is only partially successful suggests that only parts of our experience are regular. As for the rest, there one confronts a resolute kind of amorphousness – something ineradicably resistant to scientific specification. It is the noble assumption of our own scientific culture that sooner or later everything might be explained: AIDS and the problems of astrophysics, the life cycle of the snail and the origins of the universe, the coming to be and the passing away. It is not possible to contemplate this aspiration with anything but *Attaboy* on one's lips. Yet it is possible, too, that vast sections of our experience might be so very rich in information that they stay forever outside the scope of theory and remain simply what they are: unique, ineffable, insubsumable, irreducible.

Paris

I never saw any of them again – the crowd from Tacoma, I mean. Once, many years later, I received a letter from the girl into whose eyes I had fallen and then drowned. I was living in Paris at the time. I read the letter on one of those old-fashioned wooden cars that still clank and wobble their way through the Paris subways. Her sad, still presence peeped up at me from the pages by means of a neat blue script, the letters looped. All at once, it came back to me: The lowering sky and the languid sound and the somber Olympics on the far shore and the pebbly beaches and the murmurous, moldy forest with its strange mushrooms and the deep, creosote-woodsy smell of the place and that soul-stirring leotard. Around me in the moving car the life of Paris churned and bubbled. By then it was too late, of course.

It always is.

PART II

As Big as Galileo

One cold froggy fall day I found myself scurrying across the quadrangle of Columbia University in the company of a squat molecular biologist whose name I have completely forgotten – Dr. X. Rumor had it that he was in line for a Nobel Prize. We were walking with our chins down, trying to escape the vile gusting wind. X carried himself with the measured dignity of a man expecting momentarily to be flagged down by the bell captain of fate. The subject of linguistics came up. It was a time when Noam Chomsky had achieved a kind of uncanny academic incandescence. Like fluorescent light, he appeared to be everywhere at once. X stopped, looked into the sooty distance, his eyes watering immediately, and deferring to what he foolishly took to be my greater expertise, wondered aloud, with the marked air of a man discussing a subject beneath himself, whether any of the current controversies in linguistics actually made any sense.

"How big is this Chomsky?" he asked, holding his hands apart as if to measure a frankfurter.

I answered in a warm enveloping glow of inspiration.

"As big as Galileo."

"No kidding. That big?" said X, awed at last and mystified by the improbability of it all.

Status in Science

Did you imagine that within the academic world, at least, a spirit of perfect egalitarianism prevailed, with all of the sciences pretty much on the same footing and distinguished only by their separate concerns?

Well, you were wrong.

The fact of the matter is that the sciences are arranged in order of decreasing distinction. Mathematics and theoretical physics, for example, occupy a position of prominence denied to urban affairs or semiotics. Between the top and the bottom of the heap, in precisely defined niches, like so many species of sea urchins (I have measured this carefully), one finds quantum chemistry, chemistry, organic chemistry, biochemistry, molecular biology, zoology, population genetics, medicine, economics, psychology, sociology, and on a still lower level, systems analysis, advertising, cybernetics, communications, interpersonal relations, marketing, women's studies, and cinema.

Rez-de-Chausée.

Judging from the list, intellectual life seems most successful when its objects are inanimate, or, as in the case of medicine, dead or dying. Apparently there is something to the human soul that resists description in scientific or mathematical terms. We may understand the structure of the finite groups in a precise, mathematically elegant way (sort of), or form vast, impressive theories charting the progress of the universe-as-a-whole from the first flatulent Bang! to that remote but inevitable final state in which nothingness predominates again in the void; but why it is that a human being does what he does, when he does it – *Why'd he eat them nails, man?* – this remains, we are inclined instinctively to believe, a mystery as well treated by astrology (because he's a Leo), women's studies (because he's a man), EST (because he's an asshole), or Assertiveness Training (because he's a wimp), as by the sober methods of the hard sciences – hard, evidently, because their subjects are, in comparison to ourselves, sponge-rubber soft.

Dim Sum

Every human being is a three-dimensional object (four, if physicists are counting), a kind of hot localization of energy and matter and, as such, amenable to study by the methods of theoretical physics and mathematics. Pairs of politicians taken together make up a quartet. This is purely a mathematical observation. Quantum physicist and Jungian analyst, when dropped from a great height, fall at the same rate of speed, their descent unaffected by speech or creed. Physics holds sway here. We all begin things as a single nacreous and undifferentiated cell, and come to much the same end, whether from heart disease, heartache, or a simple sense that it is time to move on. In one large aspect, at least, we are such stuff as genes are made of, so much blubbery protoplasm, a cousin by means of the genetic code to every shambling and repulsive thing that lives.

For all that, most of us persist in the belief that the surface of our skins encloses somewhere a throbbing center of consciousness – unique, incommunicable, irreducible, our own, us, me. It is this impalpable stuff that comes into existence when we are born and departs when we die for those regions of the night where incorporeal souls gather to pass the time until an occasion arises to steal once again into the blessed world of matter.

Descartes said as much in his *Meditations*. Two substances predominate: Mind and Body. We are directly acquainted only with the first. What is mental is not physical. What is physical is not mental. He was unable to say, though, how the two interacted – how something intangible as a desire or a dream, moving vaguely or vigorously across a screen of consciousness (my dream, my screen), could evoke a dark firestorm of electrical activity in the crenulated folds of what is, after all, so much nervous physical jelly.

Bishop Berkeley, who remains to my mind the philosopher's philosopher, a specialist in the defense of positions that are known to be absurd, thought to get rid of the physical by arguing for the existence of thoughts, and nothing else. Such is *radical idealism*, a doctrine that

Dr. Johnson is widely thought to have refuted, chiefly by kicking a stone. Hobbes, on the other hand – am I going too fast? The history of philosophy, like Chinese food, is meant to be eaten rapidly – argued with great gusto that the universe is made only of matter. Human beings enter this scenario as complicated mechanisms, similar in kind, I suppose, to those ingenious Swiss clocks in which promptly at the quarter hour some figure emerges from the panel on the woodworks, lifts a tiny hammer, and strikes resonantly a gleaming, golden bell. Matter or Mind. The spirit or its coarse coffin of perishable flesh. Or both, as in Descartes, in uneasy alliance.

The Gnome

Some weeks later, as luck would have it, I traveled to Boston to interview Noam Chomsky for *The New York Times*. It had long been the policy of the *Times* to restrict its coverage of science pretty much to the publication of very solemn and reverential articles on the Big Bang or the latest quirks in quarks. Standing with one hopeful foot in the door, a career in journalism a matter, apparently, only of saying the right thing at the right time, I suggested to Henry Lieberman, the editor of the paper's science section, who was short and stocky and conveyed with uncanny success the impression that he found all forms of human communication largely a waste of time, the wonderful world of work that the two of us might enter together. Lieberman sat in his shirtsleeves in his dark, dismal little office, the stub of a cigar moving from one corner of his mouth to the other. When I mentioned Chomsky's name, he said "Why should we bother?" in a defiant kind of bark, his small shrewd eyes narrowing so that the bunched epicanthic folds on the top and bottom of his eyelids almost touched.

I knew more about linguistics than I did about journalism, which is to say, slightly more than nothing at all. Some twenty years before, Mrs. Crabtree, my eighth-grade teacher, had announced one dappled day that we were to learn parsing. She rose from her seat, thin as plywood, gray-haired, her long, sad, hopelessly sexless face shaped like a horseshoe, and proceeded to write a sentence on the blackboard in that flowing copperplate in which she took such robust and pathetic pride. Underneath the sentence she drew a kind of curious and disjointed graph. From the back of the room, a spitball sent into motion by Huey, the class clown, a congenitally retarded, monstrously aggressive little hoodlum, lofted in a graceful arc and landed with a dismal plop. Parsing was discreetely dropped from the curriculum. The class returned to fractions and the history of those American Indians who, in 1953 or thereabouts, were still selling real estate for absurdly low prices or drinking firewater or losing battles with a mournful whoop.

I called Chomsky from a pay phone on 42nd Street and then set off for Boston by bus.

MIT, which I reached six hours later, I remember chiefly for a style of architecture so depressing that it might have been created entirely to settle a grudge or a wager – as when an indignant and outraged husband takes a chainsaw to his half of the split-level house after the divorce. I walked for several miles in the wrong direction along a huffy-chuffy river, with that peculiar, wet, sea-smelling wind of Boston blowing fretfully in my face.

I finally found the right building. Chomsky's office, I was told by a pale, thin-armed graduate student in a white shirt, ball-point pens lined up in a row on his pocket, was in the basement somewhere, at the end of a long, gloomy tunnel. Now and again I could hear as I walked some hoarse academic signal groaning in an urgent bladderlike buzz.

Chomsky appeared shortly: A compact figure of medium height, with thick, wavy hair, parted well to the left, the main wave rising in an accountant's pompadour, heavy black glasses, rather large ears, and a flat, straight mouth, expressive in a curious way of some settled, habitual, almost prehistoric ferocity. He was sniffling profusely as we shook hands.

"I'm sorry," he said. "I have a cold."

There were books everywhere in his scruffy office, piles of paper, a decrepit and old-fashioned mechanical typewriter. Light came into the room from a small, high window and a bare electrical bulb suspended from a severe, rather artistic black cord. We sat and talked comfortably for an hour or so about language and linguistics, psychology, behaviorism, philosophers we both knew. The Korean-made tape recorder that I had bought in New York failed promptly, the tape spinning soundlessly on the reel. Chomsky was articulate without being eloquent. I now remember only one remark.

"The trouble with empiricism," Chomsky said, his voice no more than a cold-blotted murmur, "is that it lacks a fundamental concept."

I nodded, prepared to agree to anything.

"Before you can talk of *how* an organism learns anything, you have to have some idea of *what* that organism has learned, some characterization of its mature competence. This is a concept that is simply missing from empiricism."

I endeavored to convey the impression that I knew a good deal of linguistics, this despite the fact that everything I did know I knew either from hearing Chomsky talk or reading his books. If Chomsky found it disagreeable to hear his own views repeated, he did not say so, and when I found the occasion to make a point that he had made before, making it in virtually his own words, he nodded vigorously.

"Exactly," he said.

Knowing and Known

In a standing rebuke to the second law of thermodynamics, biological creatures resist for a time the tug of entropy, acquiring energy instead of dissipating it. Growing in slow stages, a human being, unlike a stone, becomes more rather than less ordered, at least for a time. Biologists, and no one else, explain this singularity in the scheme of things by arguing that living creatures do not constitute a closed system. The energy they require they derive from the sun. But the sun shines alike on the quick and the dead. Only the former hustle industriously, or take meetings at The Four Seasons, or file suit in Superior Court, or wonder petulantly where the men are. The problem, to my mind, remains open. The higher organisms – anything above a rock star – grow in what seems to be a double sense. As their mass and complexity increase, they generally get smarter as well, and come to form an increasingly refined representation of the external world, with even the dog establishing for himself, say, that voiding on the Oriental rug is not a good idea. In the end, physics does exact its awful price. This is double-entry bookkeeping with a vengeance. For most of the biological world, all that remains after death are the instructions in DNA of how to repeat the whole business from scratch. Only human beings leave lying about, in ink, pigment, stone, or silicon, some less perishable record of how for a time they saw things – their system of knowledge and belief.

Of our own lives, speaking now of just one species, half the time is spent learning the ropes, and the other half leaning against them. It is logically possible, I suppose, that everything we think we know as a result is false – a suggestion that evokes satisfying grunts of "Weird, man" from the fraternity boys in the back row whenever I present it in class. It is weirder still, given the time available, that anyone (or thing) ever comes to believe anything at all instead of exiting from the scene convinced only that life is a matter of a brief exposure to chaos. In the *Meno,* Plato put the problem of knowledge to a pliant slave by means of his own probing puppet. Step by step, Socrates pushes the thing backward, until he concludes, with the marked air of a man discovering a conclusion to which he had long since given his allegiance, that whatever might be the case elsewhere, Meno, at least, knew only what he had known. Knowledge (or true belief) is a form of recollection. This is a position that has the baffling charm of a mirror seen in a mirror, the pair of images receding in the infinite distance. But, then, Descartes also argued that in certain respects the mind came to its experiences stocked like a little trout pool with *innate* ideas – the concepts of Euclidean geometry, for example. Seeing a distorted triangle, the human intellect recovers a regular, trilateral figure. But how, unless the figure was there from the first?

In England, the contrary idea gained ground. This is interesting historical evidence that however good an argument in philosophy may happen to be, it is generally not good enough. The mind itself, Locke, Berkeley, and Hume agreed in rough measure, is largely a *formal* object; the mechanism by which it works involves only certain principles of association or coordination. What we know of matters of fact may be tracked to bright, individual bundles of sensory experience; it is the association of these experiences that allows the mind to form complex ideas. In a very famous, quiveringly alive argument, Hume concluded that even the concept of causation could be resolved into nothing more than the spatial and temporal association of events. With respect to matters of substance, the mind itself, Locke wrote, starts things off in a condition rather like a blank slate or *tabula rasa*.

Whatever the merits of British empiricism, this last

is an hypothesis with what will seem to every college instructor like overwhelming support.

Mais, je divague.

Harvard

After spending an hour with Chomsky, I visited Roger Brown at Harvard – all autumn leafiness after MIT, and bustling, bright-eyed, curiously bustless coeds. I walked over to the academic quadrangle, impressed, despite myself, with the confident elegance of the place, and consumed by the unreasonable suspicion that within minutes I would be spotted as an interloper and asked, unceremoniously, to leave.

Roger Brown turned out to be a tall, stooped, rather shambly figure, dressed comfortably in tweeds, with a warm-bath, Tom-Collins sort of personality. A psychologist by profession, he had been moved and impressed by Chomsky's work, and interested in seeing at first hand how it was that children learned their native language. In a series of experiments, he had surrounded himself with infants, and recorded their linguistic progress from day to day in an engaging, personal, and charming style. While his research had none of the acid drama of Chomsky's war of One-Against-All, it was impossible to read of his interactions with his charges without a bemused, tsk-tsking, what-some-men-do-for-a-living sense of sympathy.

We walked from his office to the faculty club, with Brown bobbing or nodding to a dozen men, the inclination of his head, I thought, indicating in some system that I could not understand the precise moral temperature of their relationship. Once in the club, we ascended a stairwell bordered by a heavy, oiled bannister, crossed through a hall on whose walls hung pictures of nineteenth-century Harvard deans, all mutton chops or heavy whiskers, and finally emerged in the dining room, which was crowded and smelled of some curious mixture of tuna fish and a deep, unknowable gravy.

A plump, rumpy, high-cheeked, brown-haired young coed from Radcliffe or somewhere popped up at our table and welcomed us both with a Chinesed "Hi" covering four distinct tones. She asked whether we would care for the special, which that day was something New Englandy and inedible – scrod or cod, baked or broiled.

"Is it any good?" I asked, my interest in the food entirely a matter of keeping our waitress's buttery skin on full view for a moment longer.

"Absolutely disgusting," she said promptly.

Lunch came and went. Brown talked lightly of linguistics and the linguists that he knew. For Chomsky he had the ungrudging respect of an intelligent man contemplating an intelligence greater than his own. I admired his graceful and unthreatened modesty. Still, he had scruples. There was something to Chomsky's personality that was, perhaps, a bit too radical. He had little feel, Brown suggested, folding his lower lip beneath his teeth to catch a stray sprout, for the continuity of science. Some of his work sounded a note of stridency. And he was not generous.

"It's curious," he said, referring to Chomsky's famous review of Skinner's *Verbal Behavior*, "that so few people have actually read the book."

I looked around the room, hoping that the subject would change itself in mid-air.

"I suppose you've had a chance to study it carefully?" Brown asked.

"Of course," I said, not even skipping a beat, the taste of tuna now oppressive on my tongue. "It really is a piece of shit."

"No, it's not," Brown said, looking pained.

Much later, I came across a used copy of Skinner's book in a stall along the banks of the Seine, and actually read a chapter or two while standing on one foot. I was right about the book, of course, but that was no excuse at all.

The Scientific Method

It is a fact: Just when a dogma or doctrine looks as if it is about to expire in the learned journals, it pops up in the textbooks, seemingly no worse for wear and glowing with ruddy good health. In looking over a college text on sociology (Frankman and Presser's *The Parameters of Social Life*, actually), I spot, right after the preface and a very long list of acknowledgements (Max Weber and Emil Durkheim, who evidently communicated by means of a Ouija Board with the authors, and their patient and long-suffering wives, who typed the manuscript in stages, four hands on one keyboard), a section devoted to the scientific method. It all seems pretty straightforward, that method, and amounts intellectually to two steps and a hop. First, the sociologist lets the facts and factoids wash over him. Step One. Then he classifies the data by means of rigorous statistical techniques. Step Two. And, finally, after immersion and classification, and a good deal of painful pondering, he frames a hypothesis, a kind of guess as to what it all means. Hence the hop. Now, there is more to the method, of course. That hypothesis must be tested. If the hypothesis is supported by the data, it becomes part of a theory. If not, the sociologist repairs to Step One. But these are the essentials.

I mention all this not merely to make fun of sociology (that too). By now, it has become commonplace among philosophers to observe that there is probably more method to golf than to science. The sleepwalking Einstein's discovery of the principles of General Relativity, for example, represented nothing more methodical than the lunge of a muscular mystical intellect toward insights that others could not discern. But the very idea of a scientific method makes contact with two concepts that are neither dead nor dying nor even indisposed. The first is the notion that any intellectual activity involves some fixed series of steps. Philosophers who are prone to pooh-pooh the idea of *the* scientific method nonetheless argue that other forms of reasoning are fixed and formal and could in principle be taught to a computer. And in its larger aspects, the very idea of a scientific *method* is linked obviously to the prevailing themes of British empiricism: Frankman and Presser's brief simply applies the empiricist's account of knowledge to the special case in which the knowledge is scientific. Many philosophers see little method in science yet count themselves empiricists.

Laws of Psychology

Theories, and not facts, are the hot breath of science, but theories without facts make for metaphysics merely. So far, I am inclined to say, so good. Science is a delicate affair, poised precariously between a world of theory and a world of fact. Yet in psychology, a theory of what? And supported by what facts? In the *Principles of Psychology*, the urbane William James argued that the subject of psychology is consciousness, and supported his claim with what amounts to an endearing shrug: What else is there?

In the ants, the bees, and the Japanese, consciousness is inescapably social, with the group itself acting as the center of self-awareness, the individuals figuring in the enterprise as bit players. Elsewhere, things are arranged in a discrete, one-to-one fashion. I am acquainted with my own consciousness, you with yours; there are, in all, two turbid centers of self-awareness, not one, and no revolting overlap either, as in a hot tub. This suggests that the subject of psychology is not public. Specialized methods of introspection are required for its analysis. On the Continent at the end of the nineteenth century, the German academic psychologist William Wundt took this advice to heart. In his laboratory, over the course of many long decades, he carried out a series of methodical, detailed, painstaking investigations into his own consciousness and that of his students. Wundt studied attention, memory, awareness; he developed a theory of the emotions and a system for the classification of various mental states. He worked tirelessly in the fashion of those nineteenth-century pedants who seemed to have occupied a temporally more spacious universe than anyone before or since. His collected writings fill almost seven volumes. But given that his approach was entirely a matter of attending closely to his *own* mind and reporting the results, his research appeared somewhat like a report by fish intended for birds (it's cool down here and very wet). Disciples, after leaving his laboratory, often remarked peevishly that when *they* looked inward, what they found was nothing like what Wundt found at all.

Noch so was!

To imagine that psychology is the science of consciousness is to seal off the subject from public inspection. But human beings live in the double world of experience *and* behavior. If consciousness remains inaccessible to science, behavior is achingly at hand. What a human being *does*, as opposed to what he thinks, believes, imagines, wishes, or feels, displaces space in the same large world that science scouts when astronomers chart the movements of the planets, or physicists compute the charge between magnetic particles.

The simple verbs of English are almost all biologically coordinated: to move, to get, to reach, to strive, to grab. This suggests, provisionally at least, the linguistic primacy of behavior – what we do, what gets done. A robust view of behavior as a matter simply of biological action no doubt seems somewhat naïve. So much of human behavior is complex. But complexity often involves simple acts heaped together, as when embezzlement is constructed from grabbing and running. A science of behavior has but two kinds of facts to study, simple and complex, and two worlds in which to embed these facts. There is the world that impinges on an organism, and the world of the organism's behavior. Cause and effect, *stimulus and response*.

The idea that psychology is nothing more than the science of behavior owes much historically to the work of the Russian physiologist Ivan Petrovich Pavlov, the He and the Hero of *Pavlovian conditioning*. Moody, self-absorbed, tense, bearded, passionate, Pavlov appears to have stepped vigorously from the pages of some elephantine nineteenth-century Russian novel in which tormented intellectuals are forever tramping about muddy country estates and worrying overmuch about their serfs – the sort of thing that the British are so adept at recasting for television. Early in his professional life, Pavlov acquired the conviction that he was meant to consecrate himself to the scientific method; toward his research – toward *research* – he adopted an attitude of innocent devoutness. Other men of his generation of Russian intellectuals, consumed by the same pointless passionate purity, became revolutionaries in a variety of gloomy causes, or, after a struggle, sank back into religious belief. Over the course of a very long, morally unblemished professional life, Pavlov trudged daily to his office, wiped his muddy felt boots at the door, and carried on, indifferent in the end to economic ruin, riot, or the Russian Revolution itself. There he would sit at his laboratory bench, furiously absorbed, until hunger or a message from his long-suffering wife prompted him to satisfy some irrelevant animal appetite for food or sleep or human warmth.

In the *reflex arc* – the hammer's mild tap followed by the knee's minor twitch – Pavlov imagined that he had discerned the very unit from which behavior itself is built. He dealt his life long in dogs, those pliant if unwilling laboratory subjects, and observed in his most famous experiment that in the presence of food dogs tend to salivate, with great ropey strands dripping from their open mouths. Food is the trigger, salivation the response. The connection between the two is a matter of empirical fact – unlearned and unmediated. The reflex arc thus makes for a behavioral atom, as given and as ineradicable as the valences of carbon, and as obvious.

The story as it continues is now a part of scientific myth. Pavlov thought to vary the stimulus that he presented his animals by ringing a bell when he offered them their food. After several trials, the animals came to *associate* the bell with the food. Salivating now in the presence of two stimuli where formerly there was only one, they soon began to salivate quite independently on hearing chimes in the distance – an elementary canine case of putting two and two together and reaching five, since the *point* of the exercise was precisely to get the poor beasts to salivate in the absence of food. What this experiment indicated, other than that dogs are easily misled, is not all that clear, but Pavlov was pleased to have elicited a measurable response from his animals. Indeed, there *is* something here that goes quite beyond the expected. The reflex arc that links in one limpid whole the salivary glands to their trigger in the nose (dogs salivate when smelling food) may well be innate, but Pavlov demonstrated something novel – a *conditioned reflex,* as he came to call it. Those dogs now salivated on hearing bells. Despite the fact that the dogs ultimately got things wrong, this would seem to count as a simple case of learning.

"In our experiments on the salivary gland," Pavlov wrote, "at first we honestly endeavored to explain our results by imagining the subjective conditions of the animal. But nothing came of it except unsuccessful controversies, and individual, personal, uncoordinated opinions. We had no alternative but to place the investigation on a purely objective basis."

Pavlov's research had an immediate and natural influence on the development of Soviet science: What better creed for concentration camps than the doctrine of behavioral plasticity? Its influence extended to the United States as well, reaching alien shores alongside the more vivid and altogether more fantastic body of Freudian doctrine. The American psychologist John Watson, an advertising executive for part of his life (it figures), championed a kind of *radical behaviorism,* and came to deny the existence of consciousness, and then the existence of Mind altogether. What was left when the mind was subtracted from behavior was simply behavior, of course, an ambitious piece of arithmetic (cognate actually to a form of division by zero) by which psychology canceled its chief abstraction.

The immense intellectual vulgarity of Watson's principles of psychology troubled a good many thoughtful observers, who tended to resolve their doubts in something like a spastic act of denial. As so often happens, Watson's scheme was superseded by a system even stricter in its commitment to behavior. Meditating at Harvard on the matter of psychology, B. F. Skinner came to the conclusion that Watson's system was simply too inflexible to account for the finely discriminated, perfectly adapted behavior of even the coarsest of creatures – pigeons, say, which Skinner unaccountably favored as laboratory subjects. The difficulty with classic stimulus-and-response theories, Skinner saw quite clearly, lay in their inability to account for behavioral novelty. Pavlov's dogs came to the experimental situation prepared by nature to salivate in front of food. Pavlov taught them to salivate on command. The stimulus changed, not the response. This schematism tied the organism to a narrowly fixed behavioral repertoire. It did what it did when it did it. Occasionally, the stimulus prompting its behavior could be altered. Left unexplained, on this account, is the emergence of the unusual – the inspired leap of intelligence that leads a dog woofishly to drop a chewed lead in its master's lap, the idea of romps and rabbits clear from the eager look in its brown eyes. In response to this difficulty (an example of the method self-applied), Skinner argued that biological organisms are by their nature disposed to try a variety of strategies in any given experimental situation. A rat placed in a maze, for example, will explore various pathways *randomly*. In this regard, the behavioral disposition of the organism is much like the biological variety of a species – something given, and hence unexplained at this level of analysis. However much the rat may be disposed to try things out by chance, a few of his strategies will prove *reinforcing* in the sense that they induce him to do what he did for a second time. It is thus, apparently, that rats do learn that when it comes to mazes the shortest distance between the starting point and a sugar pellet is by means of a fixed series of steps. Such is *operant conditioning*, what Skinner termed *selection by consequences*. The analogy to Darwin's theory of natural selection is vivid. In a series of ingenious and important experiments, Skinner demonstrated conclusively that the theory of operant conditioning explained a variety of animal acts under precisely defined laboratory conditions.

Indeed, Skinner went further in one dramatic respect. The various principles of reinforcement by which he taught pigeons to play Ping-Pong (an actual experiment) applied, Skinner argued, to animal behavior regardless of the sex or species of the organism. This was a remarkably bold, even a reckless, hypothesis. There is a great deal of difference between a great ape and a human being, still more between a human being and a pigeon. One might have expected the dynamic laws of their behavior to be different in kind as well as degree. This is to be misled, Skinner countered (or should have countered), by the superficial differences between various species. Nothing in our ordinary experience indicates that the path of a speeding bullet, the movement of the oceans under the influence of the tides, and the great swinging arcs assumed by the planets in the night sky may be explained by reference to a single force. Newton demonstrated that they arise as a response to gravity, that milky and attractive fluid bathing the cosmos in primitive models of physics. In reaching this conclusion, Newton followed a pair of magnificently simple epistemological principles: Ignore as much as possible, attend to the Essentials. In arguing for the existence of a *single* set of dynamic behavioral principles, Skinner suggested broadly that what was good enough for Newton (and physics) was surely good enough for him (and psychology).

Double Trouble

On the ordinary theory of psychology that we all apply to the day-to-day affairs of life – *He's weird, man* – it is normal to explain one psychological activity by means of another. What a man does – eating nails, say – we explain in terms of what he believes and desires:

"Why'd he eat them nails, man?"

"He thought they were nutritious."

In this way, an action is given an account by reference to certain psychological states, the first part of a convenient and familiar loop, but not one calculated to go beyond psychology. In the nature of things, we explain a man's psychological states only by an appeal to what he does:

"How'd you know he thinks them nails are nutritious?"

"Well, he's eating them, man."

The links of this loop, when followed from one point to another, suggest in outline the shadow of a larger, more general theoretical affliction. When an analysis of mental life is given in terms of mental states, it is difficult to avoid the uncharitable suspicion that precisely the capacities we need to explain we are invoking for the second time. Within Freudian psychology, this particular paradox emerges with astonishing insouciance. "It is obvious," Freud writes in discussing the common neurotic state, "that the ego is not a trustworthy or impartial agency:

> The ego is indeed the power which disowns the unconscious and has degraded it into being repressed; so how can we trust it to be fair to the unconscious? The most prominent elements in what are thus repressed are the repudiated demands of sexuality, and it is quite evident that we should never be able to guess their extent and importance from the ego's conceptions. . . . We are prepared to find that the ego's assertions will lead us astray. If we are to believe the ego, it was active at every point and itself willed and created its own symptoms. But we know that it puts up with a good deal of passivity, which it afterwards tries to disguise and gloss over. It is true that it does not always venture on such an attempt; in the symptoms of obsessional neurosis it is obliged to admit that there is some alien which is confronting it and against which it can only defend itself with difficulty.

To judge by this paragraph – it is typical – the ego is an entity capable of forming conceptions, making assertions, leading us astray, being active or passive, willing and creating symptoms, disguising something and then glossing it over, venturing, confronting and defending – terms that in ordinary contexts are applied to the *whole* of the organism of which the ego is only a part.

It is a measure of Freud's vast skill and charm that in reading the introductory lectures, we never note the logical point. But if the whole of the human personality comes to vivid life in its parts – with characters receding along the corridors of a neurosis like so many inmates – to what extent has anything been explained?

I dunno man, it's deep.

It has been this logical scruple that has long appeared

to behaviorists as a strong point in their favor. Nor has it been only in Freudian psychology that mysteriously personable entities have popped up beneath the margins of consciousness. Experimental research indicates that in integrating the data presented to it by the senses, the brain acts selectively as a *filter*, allowing only a fraction of its visual information to emerge into awareness. This suggests the intervention of some intelligent mechanism – how else might the brain discard what it does not need? What an organism intelligently does, psychologists explain by reference (among other things) to what it sees; what it sees, they explain by reference yet again to what it intelligently does. It has become fashionable among academic psychologists contemplating the decline of behaviorism to remark that with respect to Skinner, they knew it all along. However much they may now sneer at the behaviorists, "poor dumb boobs" expressing the sentiment behind the usual "inadequate to explain the manifest complexity of behavior," the argument that I have presented is not entirely trivial.

Darwinian Days

I arrived at Princeton in winter, and thought it vastly preferable to Vietnam – the alternative I faced and rejected, largely from a heartfelt desire not to see my own ass blown off. I had come to study analytic philosophy, a discipline that even then seemed sublime in the austerity of its intellectual demands: No Latin, no Greek, no long hours in the library stacks, as in medieval history, no embarrassingly difficult mathematics, surely, no science whatsoever, as far as I could tell, except for logic, which was not a science. A simple willingness to talk without pausing for breath sufficed – that and the capacity to appear deeply worried over intellectual problems that others could not discern.

I walked from the bus depot on Palmer Square through the campus to the graduate college, which lay over a hill and on the far side of a small valley. The older part of the college consisted of a rambling gothic shambles (not worth a detour, as the Michelin might have had it); off to one side, an ugly, modern red brick dormitory had been affixed to the whole thing as an afterthought. Facing north lay a broad lawn; to the south and out of sight, the campus of Princeton University. To the west, there was a wood that led ultimately to the town of Trenton, which had, unfortunately, survived a Revolutionary War battle; to the east, where the sun rose, there was the Institute for Advanced Study, occupying, so far as I could tell, ground that was widely considered consecrated.

Social life among graduate students was conducted chiefly in the commons, a suite of dusty rooms deliberately designed to create the impression that it was time for High Tea at Oxford. Nearby was an alcove with a monstrous pipe organ, which, once engaged, acquired a drunken musical life of its own and continued to moan sonorously long after anyone had ceased playing on it. Dinners were held in a long, gloomy, stone-faced hall. The bald, bushy-browed, drowsy Master interrupted his reveries with a self-conscious snort to deliver grace in Latin. We wore academic robes to dinner, took our meals at long wooden tables, twelve to a table, and addressed our undergraduate waiter as Gunga-Din.

I don't know how things were arranged elsewhere, but in philosophy, instruction was by means of the seminar. A dozen or so graduate students would meet with a professor for three uninterrupted hours. A hard-to-acquire dexterity was necessary in order simply to get through one of these classes without appearing a perfect fool. Certain great dialectical virtuosos stood out from the other graduate students. Bernie X, for example, had come to Princeton from a Brooklyn Yeshiva. He was sallow, stooped, and had yellow teeth and curious skin, which seemed several times too large for his body and functioned anatomically as a second shaggy coat – this in addition to the one he habitually wore. In class (and in life, too) he was forever lounging like a lizard in the sun of his own success, a tricky maneuver in which a subject simultaneously radiates and absorbs heat and light. If the discussion turned on a point of logic, Bernie might appeal to the history of philosophy. If the argument appeared historical, he would lean back in his chair, a look of reptilian agelessness on his features, and, speaking in a Godfather-like dry whisper, ask whoever it was that might have been addressing the seminar just what do you *mean* by such and such, an insistent intonational arpeggio being *de rigeur* in the expression of this question, with a certain peevish croak at "mean" followed by a dramatic pause.

Our attitude toward moral philosophy was a matter of cheerful indifference to experience. The philosophy of mathematics and science we studied without knowing any mathematics or science – not even the rudiments of the calculus or Newtonian mechanics – evidently on the assumption that the philosophy of a subject was superior to the subject itself. Of the classics, we read only Plato, Aristotle, Hume and Kant – just enough to spare us embarrassment should anyone ask at cocktail parties whether we knew anything about the history of philosophy. Theoretically, we were all supposed to know French and German at the very least; but toward foreign languages the department adopted an air of marked indulgence, and tended in the case of German to regard virtually any acquaintance with the old *ich-du* as evidence of complete mastery. When it came time to test this rosy thesis, it was discovered that Bernie X, who had after all learned English, was simply unable to read a passage of simple German prose, and sat looking at the text in hopeless perplexity, like a dog being asked to speak. After several fabulous flubs, Bernie was finally asked to translate a passage of pure mathematics from the German *vive voce* and managed somehow to steer his way through a handful of German words and a mass of standard symbolism ("thus", "so we see", "therefore", "finally") until he reached the end with a grateful gasp. "I did it," he said, when he had recovered his breath.

If nothing else, those endless seminars convinced almost everyone that no intellectual position was immune to analytic dissection. A few students came to Princeton prepared to argue for some rural master who had loomed large in undergraduate life. By the semester's end they concluded that what could be affirmed with success comprised a vanishing circle of light in the exhilarating darkness of analytic criticism. Academics who came to deliver papers before us took scrupulous care to so winnow what they wished to say as to stand on the very narrowest platform of declaration. Nonetheless, a few philosophers that I recall turned left when they should have turned right or simply stopped. One poor character found himself involved in a logical mistake – the affirmation of the consequent – and with his cheeks flushed with embarrassment stood facing his tormentors (us) with absolutely nothing to say and a great deal of time in which to say it.

Philosophers such as Paul Ziff, on the other hand, made out very well. I remember in particular an occasion when Ziff came to address a seminar in the philosophy of language. Standing at ease before his audience, a cigarette between his small, well-shaped, rose-bud lips, he cut a very elegant figure in his gray suit. He had acquired the valuable trick of speaking very softly in almost a murmur, and very slowly, and conveyed the impression that he was reporting the results of his immensely complicated research with the sense that he of all people knew how tentative it all was. Very carefully he drew a distinction between the gnashing of someone's teeth, and someone's gnashing his teeth. It took him well over an hour. When he had finished, and no one had been able to fault his presentation, everyone had the feeling that something substantial had been accomplished.

Force Majeure

The pivot is 1955 or thereabouts. Dwight Eisenhower yet ruled, and so did Marilyn Monroe, Konrad Adenauer, and Vladimir Nabokov. In New York, at least (I can only speak from my own experiences), Freud's image of the human soul as a haunted house held sway in that larger world whose coordinates were fixed by literary criticism, psychoanalytic journals, and that drab, dingy thing, the popular imagination. *What a man is, is what he wants*. Yet at the sullen academic center of the subject of psychology, like a pit, inaccessible to any form of light less lustrous than a government grant, behavioral psychology had hardened to an orthodoxy. *What a man is, is what he does*. Philosophers capable of splitting hairs to the molecular level, while they might have scrupled at the details of the behaviorist program, were nonetheless committed to its essential viability. Very many of them looked forward to a delicious academic partnership. The psychologists would determine the laws of behavior; the philosophers would tell them what it all meant.

In part, behavioral psychology achieved its preeminence by its rhetorical techniques. *Force Majeure*. B. F. Skinner, who was capable of writing scientific prose with a queer, disabling charm, was a master of any number of canny tricks: *avuncular exhortation*, for example. "The methods of science," he wrote, have been enormously successful wherever they have been tried. Let us then apply them to human affairs." Let us, indeed. Few of his readers observed that of the two parts of this claim, the first is false, and the second, a non sequitur.

Or there was *feigned revulsion*, to coin another phrase:

"The conception of the individual which emerges from a scientific analysis is distasteful to most of those who have been strongly affected by democratic philosophies."

Most of *those*, note, not most of us.

Or again: "When we turn to what science has to offer, however, we do not find very comforting support for the traditional western point of view. The hypothesis that man is not free is essential to the application of scientific method to the study of human behavior."

American academic psychology passed through its rhetorical high summer at a time when it seemed to many scholars that the social sciences had actually caught up to physics in their method and technique. In economics, there was game theory, which for a moment looked as if its applications might extend beyond children's finger-counting games or tic-tac-toe; elsewhere, statistical decision theory, automata theory, graph theory, network theory, mathematical sociology, cybernetics, information theory, and various theories (all absurd) of complexity and communication. Humanists might well recoil from behavioral psychology, and European philosophers continue to babble, as they have always babbled, of *angst* and dread, free will and the existential dilemma, the authentic personality, evil, goodness, and the death of God or his silence – but what of it? It wasn't scientific; it lacked the authority conferred on behavioral psychology by the ruby-eyed white rat; it was gaseous, unexperimental, self-indulgent.

And, of course, it didn't pay.

Making Out

In the fall, the campus was lambent with high color, the tall, graceful oaks and elms shedding their leaves by means of a process that involved an artistic mixture of sadness and hysteria. On Friday afternoons, an elephant troop of yellow buses would arrive at Palmer Square, disgorging, after they had come to a rumbling halt, a procession of neat, well-dressed, immaculately scrubbed, twittering coeds from Radcliffe or Vassar or Bryn Mawr, with the last bus each and every weekend reserved for teachers-in-training from Trenton State College. My own students would be there in the warm sunlight, dressed in tweed jackets, button-down shirts from J. Press, mustardy chino slacks, and scuffed white buck shoes. After an awkward shuffle, they would manage somehow to pair off with their dates. I found the whole process irresistible, and once or twice hung around the square hoping, I suppose, to be tossed a reject: *Take the one with the humungous knockers*. Occasionally, one of my students would introduce me to his date. "Uh, Dr. B'linski" – the doctorate, which I did not yet possess, a nice flattering touch to which I always responded – "this here is Dee." Cute as a bright button, dainty Dee would smile up at me, her perfectly straight and immaculately capped teeth glistening. "Pleased to meet you," she would say, inclining her body so that the faintest trace of a curtsey was somehow suggested.

Over the course of the weekend, my trusting, infinitely tender students would become progressively drunker in their eating clubs. When the buses departed Palmer Square late on Sunday evening, those scrubbed, brushed, hygienic and unavailable coeds had the sad ratty look of women tired of being mauled, and talked among themselves entirely of Can you Imagine what He asked Me to do.

No! That? Gross!

Can We Talk?

We live for the most part in a world of meaning rather than things, with the sentences that we comprehend falling about us like so much snow: stolid plumbers, their speech blooming with profanity – *I fuckin' fixed the fucker yesterday, Mr. B'linsky*; goofy newscasters mixing tragedy and trivia in a single sentence; politicians explaining how *they* didn't do it and *it* never happened or chucking it all for a slice of the commercial pizza – *I've made a choice, that's what it's all about*: Jump cut to the can of Diet Pepsi, sweating slightly, tight on Geraldine's teeth (not the legs, thank God!); sappy psychologists babbling about self-realization, self-improvement, relationships – *Before anyone can love you, you 've got to love yourself* (I do, they don't); mournful feminists – *I'm too strong for the men I meet* (too something, in any case); hick-chic novelists talking about easy women and fast foods (or the reverse); great champions of abnegation (Sufi Masters, Synanon Spokesmen, EST converts) grabbing at you in apartment corridors or the open inescapable streets – *It came to me all at once, you know. I'm an asshole, you're an asshole, we're all assholes* (one third right, at least); beady-eyed Born Again Christians explaining solemnly how Christ entered their lives (in the dead of the night generally, beating a drum, causing a good deal of commotion); streetwalkers talking that universal Esperanto – *Wannapartisweetie*; what have you. We understand them all, catching meaning somehow before it falls like brightness from the air.

Beyond the English language, there is the diversity of language itself: French, with its light lilt, and Italian, smooth as pasta, incomprehensible; German, all pomp and power, the verbs bunched at the end like a caboose; the singsong comedy of Swedish or Norwegian; Finnish with its seventeen unlearnable cases; the great Indo-European family, which includes Sanskrit, Greek and Latin; Oriental languages, forever trapped somewhere between the glottis and the thorax; strange African languages in which the sounds are made by clicking; Arabic, with its lovely script, and Hebrew; the throaty growl of Japanese; the languages of the Indian subcontinent, almost forty in all; Kurdish and Bantu, and the weird inaccessible language of the Afghan plain, with its ominously rich vocabulary for vengeance; all in all, over three thousand separate languages, uncountably many dialects; and in each, a man may express his thoughts and communicate his sentiments; in each poetry may be written; each may be learned as a man's mother tongue; and each is a world unto itself, perfect and complete.

This is what linguists often say. Whether *The Review of Astrophysics* could with ease be translated into Bantu is another story. Why human beings should require three thousand separate languages but only one digestive tract, I do not know.

Give Baby the Orange

Describing his subject, the American linguist C. F. Hockett once had occasion to remark that linguistics is a "classificatory science," like botany, perhaps, comparative anatomy, or entomology. In the case of linguistics, the object of classification is the corpus (or corpse) of a language – those utterances that an enterprising field researcher studying Cherokee or Watusi extracts from conversations overheard. The image is frankly anthropological: Linguist, tape recorder, a favonian wind over the hot and fragrant plains; in the distance, the mellow roar of lions, dusky natives, the women bare-breasted, jabbering away. To the corpus the linguist brings the same general tools as the anatomist. Their common aim is to dissect the whole into its least parts. In the case of language, the least elements are *phonemes*, which function as the smallest significant units of sound; *morphemes* are the smallest collocation of meaningful phonemes. After the morphemes come words, word classes, sentences, and sentence types. Immersed or immured in his corpus, the linguist proposed to construct a set of rigorous *dis-*

covery procedures by which the classification of a natural language could be *mechanically* achieved. With those procedures in hand, he need only present that hypothetical machine with the data and wait until the machine responded agreeably with its analysis.

This view of linguistics, and the nature of linguistic inquiry, merged amiably with behavioral psychology. The linguist studied a finite body of data. This he endeavored to decompose into universal constituents by purely mechanical means. His was a science and a subject uncorrupted by an appeal to meaning or intuition, and untouched by mental entities. Only as a coincidence was the linguist prepared to acknowledge that the corpus that he studied was a part of a living language and formed, as such, a system intended to express thoughts or convey sentiments or record facts. Behavioral psychologists assumed that this very same finite corpus could be studied as part of an extended pattern of stimulus and response – what philosophers often called *dispositions to verbal behavior*. In a study such as Leonard Bloomfield's *Language*, these points of view came explicitly to coincide. Thus Bloomfield wrote that "a language presents itself to us, at any one moment, as a stable structure of lexical and grammatical habits." It is quite true, Bloomfield noted as an aching afterthought, that a language may go beyond any of its finite samples; those speech forms, or sentences, that are novel – the afterthought now reaches its anticlimactic aftermath – a speaker produces by quite trifling substitutions in speech forms that are not. In this way, Bloomfield argued, "Give Annie the orange" is formed by analogy to "Give Baby the orange." I draw a curtain of quiet charity over other examples.

The accord between linguistics and psychology took place at a time when scholars in both camps were residing in sunny Sicily, where the women are cocoa-skinned, the temperatures warm. Now, of course, the structural linguists and behavioral psychologists find themselves pretty much living in Lapland. The snows are ten feet deep and the women are shaped like seals. But I am speaking of then.

Chomsky Ascendant

One balmy evening, Noam Chomsky came to Princeton to deliver the first of his Christian Gauss Lectures. At the time, Chomsky had not quite become a familiar academic commodity. Among philosophers he was known vaguely (in the fashion of those advertisements in which *RonRico Rum* would be mistaken for an Argentinian Tango dancer); the prevailing assumption was that no matter what he intended to say, we would slice him up like a salami.

The twilight was mild, with an outrageous sunset just beginning to shape itself in the distance. Although the wisteria had long since faded, the air was fragrant. As the evening deepened, I could hear a night insect delivering itself of a few last heartfelt and regretful chirps, the thought of winter and wetness, no doubt, uppermost on its mind.

After a while, we all trooped into Firestone library. There was the usual academic babble, an atmosphere of excitement.

Shortly after eight, Chomsky arrived. He was surrounded by a phalanx of young men, like a Latin American dictator, and moved to his seat on the stage with a stride that managed wonderfully to suggest that both his time and his temper were very short. He was introduced by some senior professor straining after the usual academic air of self-effacement and chuckling good humor. The room quieted.

Chomsky spoke softly, without notes, almost at a conversational pace. Had we expected the usual? A series of carefully hedged remarks? What we got was from the first like hot thunder.

Behavioral psychology, he began, is either vacuous – Skinner – or false – Hull.

Chomsky had a trick of raising his hands to chest height and then bending his wrists backward so that his palms were displayed to the audience in a clumsy but terribly effective gesture of contempt.

A natural language cannot be understood in terms of anything like habits or dispositions to verbal behavior. In fact, the entire notion of a disposition (to anything) is every bit as unclear as any of the concepts that in Ryle or Wittgenstein it is meant to replace – meaning, for example.

At this point a late-summer fly entered the room from the open window, and with a fine disregard for the proprieties began to circle Chomsky's head. Without ceasing to speak, Chomsky commenced a serpentine head wagging timed perfectly to keep the fly at the furthest focus of an ellipse.

In any event, he went on, the thunder coming in long lapping rolls as the rest of us were dying for him to swat that fly, every native speaker of a natural language is capable of producing and understanding infinitely many sentences that he has never heard or spoken before.

Was such a thing true? Evidently so. This is simply a *fact*, Chomsky said, catching the question from the air and then crushing it.

Tiring of linguistics, and everything else, I suppose, that fly took off for the windowpane, where it continued to buzz irritably.

At the back of the room, the philosophers crossed their legs and creased their faces as if they were unsure whether what they were hearing was being said in earnest.

And then those famous sentences: *The shooting of the*

hunters disturbed them, or *She liked her cooking*. Note, said Chomsky, that the sentences are ambiguous.

So they were.

But no word in the sentences is ambiguous.

Well, yes, true again. (But how is this *known*?)

More hot thunder, large claims, a dozen references ticked off on the fingers of two hands, the last recycled on the thumb and forefinger, double-digit inflation; a shrug, a wave, wrists bent, palms exposed – the step before a slap.

Alarming examples followed from all over the place: psychology, ethnology, automata theory, solid-state physics. It was queer how complicated English grammar could be.

For the philosophers fate had reserved a special unkindness. The chief intellectual problem of linguistics, Chomsky said, was to determine how a child acquires its native language. Toward the behaviorist thesis that language-learning is a matter of conditioning, the great analytic philosophers – Ryle, Wittgenstein, Quine – had responded with warm effusiveness, the two sides to the communion perfectly in accord with respect to the doctrine that somehow or other the human languages are *learned*. This, Chomsky said bluntly, is a matter merely of dogma. Behind behaviorism – the mistake in practice – there is empiricism – the mistake in principle.

It went on for an hour, an effortless display of intellectual mastery, ranging over half a dozen disciplines, breezy almost, but superbly controlled, a superior intelligence, utterly at ease, indifferent to criticism.

Some salami.

Loitering at Infinity

In any subject, it is useful to know, even in a rough and ready way, how many items – sentences, say –

one needs to study. To the extent that the stock of available things is finite, the physical sciences describe a bounded world, with the whole of the cosmos, stars as well as starlets, made up of the elementary particles arrayed in different ways. But it is one of the paradoxes of mathematical physics that in order to describe a finite world, physicists must employ infinitely many real numbers. This is a circumstance that suggests the inescapability of the infinite in one form or another.

In going past the finite to loiter at infinity, the intellect is under the control of a number of clashing metaphors. When the eye is imaginatively projected into physical space, a sense of pure *boundlessness* is the result; this, I suspect, is the basis of accounts that mystics offer of the voluptuous swooning of their souls. The image is one in which something very much like light is moving within an infinite volume of space. Physicists and mathematicians have argued, however, that like the surface of a sphere the universe may be finite but essentially unbounded. Beyond It, there is Nothing. Racing around such a surface, a ray of light is bound sooner or later to catch up with itself, a process that suggests more the movements of a gerbil (on a treadmill) than the motions of the soul. In picturing this, of course, I seem somehow to have peeped behind the Back of the Beyond, so that I find myself surveying the whole business from a luxuriant extra dimension. The whole thing is very mysterious.

Arithmetic provides an altogether more prosaic notion of the infinite. The natural numbers rise to consciousness like an insomniac's shaggy but insubstantial sheep and then fade quickly from sight. There are mathematicians who argue (or believe) that the numbers form a completed totality – something seen at once in all its infinite extent by an immensely specialized organ of mathematical sight. The rest of us, knowing a handful of numbers on a first name basis, trust to addition to keep the sequence going. In this way, the infinite is explained by iteration – the hypothesis that we could go on *forever* in counting, if only *we* could go on forever. The hope behind this hypothesis is just that the appearance of one infinitely large object might be explained by the invocation of another infinitely extended process. Yet for all that, no act of computation ever goes beyond the finite. Our memories are bounded; what we can see (the natural numbers) is not entirely explained by what we can do (count).

Under the Aspect of Syntax

Like so much else that Noam Chomsky has written, *Syntactic Structures* begins with a bang:

> Syntax is the study of the principles and processes by which sentences are constructed in particular languages. Syntactic investigation of a given language has as its goal the construction of a grammar that can be viewed as a device of some sort for producing the sentences of the language under analysis. . . . The fundamental aim in the linguistic analysis of a language . . . is to separate the grammatical sequences . . . from the ungrammatical sequences, and to study the structure of the grammatical sequences.

In later years, Chomsky was to argue that this program in linguistics – *his* program – had been anticipated by logicians of the Port Royal School, by seventeenth-century rationalists, by Descartes, by Leibnitz, and by Wilhelm von Humboldt. The truth of the matter is that like all great books, *Syntactic Structures* managed to create its own predecessors. To structural linguists and analytic philosophers, who were understandably unenthusiastic about a revolution in which they played no part, Chomsky's confidently expressed conviction that they had got almost everything wrong had the ominous disabling force said to be characteristic of a blow in Kung Fu. The structuralists, and behavioral psychologists, too, had thought of linguistics as largely an exercise in the finite classification of a system of speech. From the first, Chomsky thought of linguistics in terms of an alien concept – a device that might generate all and only the sentences of a natural language. This was an idea that could barely be expressed in traditional terms.

All this requires some stage-setting. (That stage is due to get pretty crowded shortly.)

Sight, of all the senses, seems the most finely discriminated and the most precise. It is curious, then, that we express ourselves instinctively in sound, but a language is most alive when it is spoken (and not read): I talk, you listen, the best of all possible worlds, or the reverse, and in any case, the only direct way short of gesture that two souls may be placed in contiguity. As it is spoken, a language conveys a distracting aspect of inevitability and indivisibility: What we say fits what we mean (or need) to say; what gets said seems as resistant to dissection as the breath on which it is carried. At the very center of the soul, my own, at any rate, thoughts simply arrive in consciousness with a sudden whoop and a whoosh. Speaking must, therefore, involve a mechanism by which a cloud of consciousness is systematically mapped to the linear sequences of a natural language. From time to time, however, some sputtering irregularity breaks the flow from thought to speech – embarrassment, awkwardness, a word that drops suddenly from sight, its replacement impossible to find. The whole business then stands revealed as a system that works on various levels. Going down, a sentence is separated into its constituents: nouns, noun phrases, verbs, copulas, adjectives, adverbs, determiners. On still a lower level, there are words, with letters appearing later still, an artifact of theory. Going up, the procedure is reversed. A sentence is constructed from its parts.

A natural language is composed of words. Of these,

there are only finitely many – perhaps 600,000 in English. Behavioral psychologists, of course, had argued that language is much a matter of habit; speech, a question of stimulus and response. Habits and responses are finite objects. And, for the native speaker of English, being engulfed in the spoken sounds of his native language *is* rather like slipping into a warm bath in its mixture of intimacy and familiarity. If anything, this image does call habit to mind, and suggests that we carry the whole of a language quite conveniently with us in the form of a few well understood and well loved staples – the words that we *know*, the sentences that we have *learned*. The fact that so much of what we hear is entirely novel injects an icy spurt into this view of things. Virtually every sentence that we encounter is new, at least in the syntactic arrangement of its elements, however much the thought that it expresses may be trite or trivial. In any record of discourse, each sentence struggles up from the page manfully proclaiming its individuality. Human speech is *creative*, a source of endless novelty, as varied, or so it would seem, as life itself.

This linguists before Chomsky knew. As so often happens, they filed the fact and forgot it. It was Chomsky who paused before the sheer size of a natural language and stood puzzled.

A natural language is an instrument for the generation of diversity (and a curious example in which natural selection has apparently overshot itself, arriving at the exotic instead of the merely efficient). The particular pattern by which relatively few words find themselves forming a fantastic sentential panorama is one that reappears after a fashion in biology itself. The diversity of life is much a matter in which a single stock of genetic words undergoes an endless elaboration – appearing now as a cuttlefish and later as a linguist. The pattern reappears, too, in arithmetic or set theory. Using only the empty set – a language in which only one word figures – the mathematician is yet capable of generating the whole of elementary arithmetic. Does this imply by analogy that the sentences of a natural language make up a class that is infinitely large? Not quite. Not yet. There are well over 10,000 nouns in English, and perhaps 10,000 verbs, and thus at least 100,000,000 simple sentences in which a noun is followed by a verb: *John boogies*, for example. The novelty of a natural language may be nothing more than a reflection of its large but finite size.

But precisely the same argument might be used to generate the implausible conclusion that there are only finitely many natural numbers. What gives pause in the case of arithmetic is just the fact that larger numbers may be constructed from smaller numbers, the operation proceeding forever. The same observation carries over to the case of language. So far as I know, my experience being limited to French and German (and that Latin that is now lost), every language admits of certain operations whereby larger sentences may be built from smaller sentences. There is conjunction, for example, and disjunction – the most obvious of cases. Negation is another example, turning an affirmative into a longer negative sentence. Why stop? Such devices, once noticed, multiply alarmingly.

A natural language, Chomsky argued, contains an infinite number of sentences because sentences are constructed at least in part by *recursive* mechanisms. Only the uncharitable observed that Chomsky also argued that those mechanisms are recursive just because a natural language contains an infinite number of sentences.

To take seriously the idea that a living language is unbounded is to entertain first doubts about behavioral psychology. Between what a man hears, the behaviorists had argued, and what he says, there is nothing at all. The system consists of but two parts: What goes in and What comes out. Yet every normal speaker of English *is* able to recognize and produce sentences that he has neither heard nor spoken before. The behaviorists' schema falls short by one third. Between what a man hears, and what he says, there is – there *must* be – an object of uncertain powers, a device of some sort, a set of rules or a set of states, but *something*, in any case. Why, then, should the psychologist or linguist attend to the contingencies of *behavior*? The proper object of their attention is not what a human being does (babble), but what he *can* do (babble *on*). The subject matter of psychology is that and everything else.

With this argument, modern linguistics and behavioral psychology part company in bad blood. But the adumbration of Chomsky's program served to alienate the old-fashioned linguist as well as the psychologist. That trickle of bad blood is now a torrent. It is this second step that takes some seeing.

My argument thus far – I have acquired the thing by morganatic means – has established only that the human brain requires a computational crutch in order to generate infinitely many objects. This is an observation to which many linguists might have given their indifferent assent. After all, grammarians since time immemorial have spoken of rules; the subject is their stock-in-trade. How so, and in what, the revolution?

In learning English, the baffled foreigner is taught that while *I* go, *he* went. The advice offered him by the traditional grammarian is largely local. Yet in starting with just four words (*I, go, he, went*), I can generate sixteen two-word sentences, and infinitely many sentences of arbitrary length. Thirteen of the two-word sentences are ungrammatical; the longer sentences, for the most part, are sheer gibberish (*I go, go, go, . . .*). Evidently, those exceptions touted by tame grammarians make sense only when understood against the background provided by a set of productive rules. The rules themselves they left pretty much in the dark. Indeed, in explaining *how* exceptions arise, the traditional grammarian assumed that the rules against which the exceptions are contrasted are antecedently understood, and thus took for granted what needs most to be explained.

Whatever the details of the dispute between the old and the new, the complexity of English grammar served yet again to strengthen the generative grammarians' case against the behaviorists. For many years, psychologists spoke of language as if it were somehow a system taught at infancy – this by means of helpful hints, systematic rewards, or outright threat. In fact, human beings acquire their native language without notable effort; they pick up what they need to know from data that is degenerate (a nice phrase, that), incomplete, fragmentary, inconclusive. In running a maze (life, for example), a man performs little better than a rat. Both species apparently adopt much the same strategy in their misery: a clumsy random search. In speaking a language, however, human beings display the sort of mastery that is characteristic of biological impulse and control: Getting Food and Getting Sleep and Getting Laid.

Chomsky's violent reorganization of linguistics – I pass now to the Maugham-magisterial summing up – thus proceeded stepwise in stages. In a memorable passage, Chomsky articulated the whole of his vibrant vision:

> Linguistic theory is concerned primarily with an ideal speaker-listener, in a completely homogeneous speech-community, who knows his language perfectly. . . . The problem for the linguist, as well as the child learning language, is to determine from the data of performance, the underlying system of rules that has been mastered by the speaker-listener, and that he puts to use in actual performance. . . . Linguistic theory is mentalistic, since it is concerned with discovering a mental reality underlying actual behavior. Obviously, use of language or hypothetical dispositions to respond may provide evidence as to the nature of this mental reality, but surely cannot constitute the actual subject matter of linguistics, if this is to be a serious discipline.

All of the great themes are here in this quotation, and may be summarized in an exchange. In place of the structuralist's corpus, an *infinite list of sentences;* in place of discovery procedures, *generation;* in place of classification, *grammar;* in place of performance, *competence;* in place of behavior, *mind.* To pivot on this axis between the old and the new is not merely to see language in a fresh light. It is also to see the aims and claims of both structural linguistics and behavioral psychology fall into the abysmal academic darkness of utter irrelevance.

Sour Grapes

In preparing my article for *The New York Times,* it occurred to me that perhaps I should speak to at least one of Chomsky's unhappy critics. I knew that C. F. Hockett taught at Cornell. I called the information operator in Ithaca from Cambridge. After the briefest of pauses during which the line went dead, an obviously mechanical voice intoned that the number *is,* and then stopped dramatically, as if the machine regarded finding a number as a minor miracle and was eager to share its good fortune around like communion wine. The strangely amiable party on the other end of the next call, although he was named C. F. Hockett (Yup), insisted perversely that he had never heard of Chomsky and knew nothing of linguistics (Nope), but would have gone on nonetheless, like a character in a novel by Dashell Hammett understandably eager to get out of those wooden pages, had I not banged the receiver down on its obliging hook just as this irritating, shadowy Hockett appeared to be warming up for a nice comfy conversation.

When I finally reached the right Hockett, the two Ithaca Hocketts evidently existing in one of those quantum mechanical states in which telephone calls and identities are gaily batted around, I said that I had a few questions to ask about linguistics.

At the other end of the line there was now only an uninterested silence.

"I'm with *The New York Times*," I added.

"Oh, *really*?" Hockett said with a sudden eruption of cheerfulness. "That's different. I thought you might have been a student or something."

"No, nothing like that at all."

"I do hope that you're going to do a *bal*anced piece," Hockett said. "Focus on the discipline and not the people in the limelight. That sort of thing."

I confessed that actually the piece was about Chomsky.

"Ah," Hockett said.

I asked what he thought of the generative grammarians.

"Well," he said – it was a very glum, meditative, disappointed "well" – "they're certainly riding high. But it's all so shallow. The fact is, these people really don't know very much about linguistics at all. I mean Chomsky doesn't even *speak* anything but English."

I asked whether he saw anything at all worthwhile in the program that Chomsky had initiated.

There was a pregnant pause on the line, which hummed vibrantly throughout our discussion. I stood with a roll of dimes bunched in my fist. The undulating tin square of the floor was covered with thin, wispy streaks of mud. Hockett plainly regarded my invitation to comment favorably about Chomsky with the markedly reluctant enthusiasm of a condemned man asked to say of his executioner that he was, at least, a good provider.

"Well," he finally said for the second time, in a sad, indifferent voice, the sound seeming to shrink, "they seem to have discovered a lot of little facts about English grammar."

Sunny Days at Stanford

Stanford University sprawls over a valley that is still sunny despite the smog; indeed, the pollution itself, when mixed with the faintly salty air from the Pacific Ocean, lends to the ambient atmosphere a touch of lascivious pink and genital-plum purple that it would not otherwise possess. To the rear of the university are the hills of the coast range, where cougars still roar at night (a frightening, thrilling sound, actually); in the far foreground, the San Francisco Bay.

The campus itself is composed of a series of amusing public squares. In one, there is a little wading pool with bright sea-green water; on its border, then and forever, a hugely pregnant young mother, a toddler, and a spaniel with long ears trying frantically to get into the pool, his back paws scrabbling on the concrete wall, and then, having gotten thoroughly wet, trying just as frantically to get out in order to shake himself in the sunlight. In another square is a curious glass-enclosed clock made of brass. Between the hours, it simply stands there, silent as the Spaniards who first came to California, but promptly on the hour it leaps into confident life and commences to whirr and chime and madly peal forth bell songs – snatches from grand opera, Gregorian Chant, Broadway show tunes. There is a bookstore, and a library, and a good many bright green lawns, and a student union, and a group of squat Taco Bell structures made of sandstone, with red Spanish tile on their roofs. In these, the various academic departments are housed. An immense church is located near the center of the campus. On its walls, the colors glowing luridly in the sunlight, is a Mexican mural of the sort that was briefly popular in urban ghettos – some sort of weak solution of socialist realism. I forget all the horrible details, but I do seem to remember an iridescent Christ, surrounded by his disciplines in flowing mauve robes, looking for all the world as if they were jointly about to disrobe and go surfing. Directly above this crowd, there is a list of famous names – Socrates, Moses, Galileo – with a few odd clunkers thrown in for good measure – Leland Stanford, Jr., who yet reposes in some campus mausoleum, Booker T. Washington, a certain General Pasquale.

I had come to Stanford from Princeton University and loved the place at once. The girls, it is true, were less pretty than I had hoped, and dressed, all of them, in white tennis shorts, which revealed their tanned and muscular but hopelessly short and unshapely limbs; but to pass from the east to the west coast is to experience at once an overwhelming suspicion that there are, after all, better things to do with one's time than work.

Sometime in the summer – it might have been December for all that the seasons really change – Noam Chomsky came to the campus to debate Patrick Suppes. At the time, Suppes had achieved a triple of professorship in statistics, psychology, and philosophy. He was gregarious and bouncy, Suppes was, with a five-sided face, dark hair, and pebbly skin. During class or in his office, he wore shapeless jackets and bow ties, pleated pants, and shoes meant to conquer adversity, but at noon he would emerge from his chambers like a sunburst, prepared for tennis in spotless white shorts and a radiant polo shirt, his racket resting on his shoulder. I never got to play him, but assumed from his bobbing and profoundly ungraceful walk that he was a mediocre athlete; colleagues told me that he generally managed somehow to return everything that came his way and won every one of his matches simply by outlasting his opponents.

However he may have played tennis, it was at conferences and seminars that Suppes simply shone. Professor Blodgett, let us imagine, has just concluded a droning and incomprehensible talk on his speciality,

Blodgett Lattices, which are algebraic objects known to Blodgett and perhaps three other mathematicians in the western world. With embarrassment and a hideous sense of wasted time hanging in the air, Suppes, and only Suppes, would swivel in his seat, his face creased by a smile suggesting rapt interest in the proceedings, and after carrying on an effort to organize and marshal his thoughts (an academic Foch, surely), he would rise, thrust a hand in a baggy pocket, and then ask Blodgett, who by now was looking Suppes-ward with an expression of canine gratitude for the attention, a question of large, general, and perfect irrelevance.

That's our Patrick, we always thought: *Macht alles.*

•••

The debate was held in a small, elegant amphitheater. I came early. Everyone from the philosophy department trooped in. There was David Nivison, who was tall and reed thin; several superannuated characters, all jowls and baffled beagled eyes, greeting each other with low, slow, throaty woofs; Donald Davidson, a neat, ascetic, self-contained man of perhaps forty-five, with splendid blue arctic eyes, a symmetric, finely-made face – thin nose, austere lips – and an air of ineffable personal superiority; the usual run of undergraduates; and an incredibly lovely, long-legged graduate student, with the brush-soft name of Zoe, a former cocktail waitress, I think, who had conceived an inexplicable passion for analytic philosophy at some point between a gin rickey and a proposition.

Suppes spoke first. He strode up and down upon the stage, flapping his arms; gradually his bow tie became skewed. His thesis, I recall, was essentially that there is no analyzing language without the theory of probability. He deplored the lack of statistical rigor in Chomsky's work. He felt that, perhaps, Chomsky was somewhat naïve in his approach (toward anything); and that, in any case, things were either simpler or more complex (I cannot remember which) than Chomsky had supposed.

As Suppes talked, his (infrequent) pauses for breath elicited a kind of sympathetic exhalation from the audience, so that from time to time everyone appeared to be wheezing together, like a concertina. As far as I could tell, what Suppes had to say made perfect sense; I sighed along with the rest of the crowd. Language is a mechanism by which information is communicated; information is what gets analyzed in the theory of information; and the theory of information is itself expressed by means of probabilistic concepts. Claude Shannon, for example, had argued (in his original paper, in fact) that the regularities of English might reflect nothing more than the statistical frequency of its elements. In a series of fascinating experiments, Shannon demonstrated that a stochastic device might actually generate prose paragraphs of a vaguely English-like sort.

With twenty-six letters and a space all varying *independently*, a representative sample appears as mere gibberish:

> XFORMYMLY RHSUSIEESY THWOPELS SHJTOSQOSP SPSJFOST AOQKRPST SKTPOST SJTOSTL WJTROA SJOTPQL SJOTOLEPTPD.

But when triplets of letters are assigned their normal English frequencies, the difference is striking:

> IN NO IST LAT WHEY CRATICT FROURE BIRS GROCID PONDENOME OF DEMONSTURES OF THE REPTAGIN IS REGOACTIONA OF CRE.

Even more striking is the result obtained when words, and not letters, are varied according to the likelihood of their occurrence:

> ROAD IN THE COUNTRY WAS INSANE ESPECIALLY IN THE DREARY ROOMS WHERE THEY HAVE SOME BOOKS TO BUY FOR STUDYING GREEK.

This sentence has a nice touch and might have been written by Virginia Woolf.

Suppes drew no firm conclusions from Shannon's work, but he suggested that only a man who knew both linguistics and statistics could discern the relationship between the two disciplines. Then he spoke of finite-state automata and Markov processes. The blackboard was now covered with diagrams and arrows.

•••

In English, words go over to sentences from left to right, one step at a time. In Hebrew and Arabic, it is the other way round, and Japanese, I gather, moves from top to bottom, the vertical motion, I would imagine, reflecting the fact that these people are always bowing to one another. To many theorists – C. F. Hockett, for example – it appeared that the construction of an English sentence was essentially a two-part process. An initial word is chosen – perhaps at random. Subsequent words are then fixed by the words already chosen, so that the overall effect is like that created by a cascade. Thus suppose that just four word classes are under consideration: (THE, THOSE), (OLD), (MAN, MEN), and (COME, CAME). A machine occupying one of a finite number of internal states is now invoked. At the commencement of its labors, the machine is in an *initial* state; at the end, in a *final* state. As it proceeds, the machine dutifully contemplates each class of symbols in turn, with one state assigned to each word. This makes for seven states in all. Beginning at the left, the machine moves to the right, picking up a word from each class. Its behavior is under the control of very simple grammatical principles. Having chosen THOSE as its initial symbol, say, the machine cannot next choose MAN. This machine generates only a few grammatical sentences of English: THE OLD MAN CAME, THE OLD MEN COME, THOSE OLD MEN COME, THOSE OLD MEN CAME.

In this simple-minded scheme, the machine passes only once through each class of words. It is thus a device devoid of infinite powers, and useless as a model of human linguistic competence. But suppose that the thing is allowed to loop. On contemplating OLD, the machine has the option either of proceeding onward, or returning to OLD again. It is capable now of producing in dreary succession sentences such as THE MAN CAME, THE OLD MAN CAME, THE OLD, OLD MAN CAME – this to the very end of time.

Such is a *finite state automaton* – an abstract device taking inputs (words) to outputs (sentences) by means of a finite set of states. Now, as I have described things, the transition from state to state is pretty much fixed from the first. What the machine does, it does because it must. Those state transitions, however, may be mediated by transition probabilities. These amount to nothing more than a general scheme of conditional probabilities defined over states. The machine chooses either THE or THOSE, let us imagine, by flipping a coin. Having chosen THE the machine has moved from its initial state of indifference. The probability that it will move next to any subsequent state is determined by the state it now occupies. If there is a 70% chance that having chosen THE and OLD it will now choose MAN and not MEN, then on three occasions out of ten, the machine will follow THE and OLD with MEN and not MAN.

With transition probabilities defined, a finite state automaton takes on an elegant secondary identity as a finite state *Markov Process*.

The idea that a natural language might be modeled

by a finite state Markov process is by no means absurd. The mathematical details to the theory are well known, and just complex enough to afford the linguist the pleasurable sense that he is making use of material that his collegues studying Old English, say, would find incomprehensible. The mathematicians having done the disagreeable work, the linguist proposed simply to expropriate the results, a wonderful example of academic imperialism. At the same time, it does appear that what I *will* say has something, at least, to do with what I *have* said. The depiction of language as a Markov process fits nicely with its description as a matter of stimulus and response. A subject who has learned to recognize the sentence THE OLD MAN CAME has thus passed through four distinct stages. Choosing THE at random, he has been taught to choose the very next word on the basis of his initial choice. His two choices having been made, his next choice is determined by the cumulative record of his choices, until, finally, the subject exits from this particular maze at the full stop of a period – the end of the sentence, the end of the line. Those behavioral psychologists who could follow an elementary mathematical argument (the minority, of course) thus tendered the linguist an alert round of applause. Two separate sets of hands may now be heard to clap.

●●●

Throughout all this – there were by now a great many diagrams on the blackboard – Chomsky sat quietly, imperturbable. When his turn came, he spoke in his usual matter-of-fact manner, hardly pacing. What Suppes had to say was nonsense. Information theory is an interesting if minor discipline; its methods shed no light whatsoever on the problems of linguistics. The grammatical sentences of a natural language cannot be approximated by any method that investigates the frequency of certain English words or letters. The odds that I might produce a given English sentence, having already produced certain other sentences, are zero.

Sitting off to one side of the stage, Suppes, on hearing this, furrowed his eyebrows together as if to suggest he were being subjected to a very great absurdity.

There are infinitely many sentences in English, Chomsky continued, and consequently infinitely many English sentences that have not been uttered. On any scheme of conditional probability, *these* sentences must be assigned very low odds. If what I will say is a matter of what I have said, my chances of saying something I have *never* said – I am adding a little luster to the argument – are the same as the odds that I might suddenly conclude my thoughts with a sentence drawn from the Japanese.

And those finite state automata on which Suppes had lavished so much attention at the blackboard? Chomsky turned his palms toward the audience as if to suggest that he was baffled by the obdurate persistence of human stupidity. We *know,* he said, that English is simply not a finite state language. He turned to face the blackboard and in large firm capitals wrote the sentence THE MEN, WHO WERE PLAYING POKER, CAME EARLY. The word MEN and CAME he connected by means of a loop. Then he stepped back from the board, keeping his body inclined so that it was displayed toward the audience at an angle. There is, he said, a definite grammatical relationship between MEN and CAME. This relationship a finite state automata cannot express – the machine itself is allowed only to move from word to word in a linear sequence. Then, as if he were finally amused by the obviousness of it all, Chomsky hunched his shoulders and smiled one of those queer Alfred E. Newman grins (teeth slightly gapped, ears jugged) that are occasionally the affliction of the most serious of men.

After the talks, there was a reception. Suppes had little to say. The philosophers stood by the punchbowl and suggested to one another that everything that they had heard was either trivially true or truly trivial. Chomsky paused by the canapes, nodding severely to the graduate students who had clustered around him, and then strode abruptly from the room and vanished into the night.

Wordsmith

In writing for *The New York Times,* Henry Lieberman urged, keep in mind a hypothetical reader with but a twelfth-grade education: "Write bright," he said. I prepared a long, turgid, thoroughly inaccessible account of my interview with Chomsky, and when Lieberman told me brusquely that the thing was unreadable (the perfect truth), I answered uncharitably that Providence had obviously not intended that I write for illiterates. Lieberman looked up at me, his eyebrows almost meeting, his cigar at the dead center of his mouth. "You ain't gonna write for the *Times* either," he said. I wrote Chomsky to explain the situation. The paper, I suggested broadly, with a fine disregard for detail, persisted in treating the revolution in linguistics with bovine indifference. The fault was theirs and not mine. I had done my best. Chomsky wrote back promptly. "Dear David," his letter began handsomely. "Dear Noam," I wrote back, adding to my letter, familiar now to the point of vulgarity, a suggestion that *he* might profit from reading one of *my* papers on mathematical systems theory.

How curious that of all the emotions only embarrassment and not love may be inconveniently recalled in all of its overpowering immediacy.

Sentences from Symbols

In speaking and understanding a natural language, and in doing much besides, human beings apparently act in conformity with certain immensely complicated rules. Chomsky defined a grammar as a device that might generate the sentences of a natural language, and argued for its existence by an appeal to its utility. Without such a device, the linguist could not completely describe the whole of a natural language. But the device that the linguist requires is required, too, by the rest of us simply to keep on talking (or trucking).

The sentences of a natural language make up an infinitely large set. This is the output, to revert to machine-like metaphors, that the linguist gets from the grammar. What he gives to the grammar by way of input is, among other things, a finite stock of words. Now the finite state automaton serves to generate sentences from words *directly*; given an assortment of words, the machine proceeds in stages to the construction of a sentence. It is this simplicity in conception that accounts for its uselessness. The *phrase structure grammars*, by way of contrast, correspond in the abstract to a form of parsing. (Mrs. Crabtree now rises wraithlike before my eyes.) By means of a set of rules, stray sentences are decomposed into their ultimate parts; with the grammar run backwards (or upside down), those parts are reassembled into sentences.

I now proceed to a prim pedagogical example. Say that among the elements of a finite alphabet are the following symbols: S (Sentence), NP (Noun Phrase), VP (Verb Phrase), T (Determiner), and N (Noun). These correspond to abstract grammatical categories. In addition, the alphabet contains a *lexicon* or list of ordinary English words. A phrase structure grammar defined over these symbols consists of a set of six rules:

1) Sentence → NP + VP
2) NP → T + N
3) VP → Verb + NP
4) T → The
5) N → man, ball, etc., etc.
6) Verb → hit, took, etc., etc.

The arrow signifies that each element to the left of a line may be rewritten as the element or elements to the right of the line. These rules may be used to generate an English sentence by means of a *derivation* (a concept destined to make any number of appearances in these pages):

1) Sentence	
2) NP + VP	By Rule 1
3) T + N + VP	By Rule 2
4) T + N + Verb + NP	By Rule 3
5) the + N + Verb + NP	By Rule 4
6) the + man + Verb + NP	By Rule 5
7) the + man + hit +NP	By Rule 6
8) the + man + hit + T + N	By Rule 2
9) the + man + hit + the + N	By Rule 4
10) the + man + hit + the + ball	By Rule 5

In this exercise the grammatical categories appear early on; the words themselves make an appearance late in the derivation. A derivation is *complete* when every symbol on a given line is made up of words. It is then that a hypothetical speaker knows that S – a sentence – is grammatical.

A grammar organized thus differs from a finite-state grammar or even a Markov Process in the prominence it accords the abstract grammatical elements of its alphabet. These categories, it bears noting, are not themselves a part of the langauge – not English, in any case. Objects such as VP do not appear directly in the silken stream of speech. It is the appearance of these abstract categories that makes possible multiple representations of a given sentence. As one moves downward, each line in a derivation uncovers progressively more structure, so that the overall effect is rather like that achieved by watching a flower unfold with the exquisite languor that is caught only by time-lapse photography.

Phrase structure grammars constitute an example merely. I mention them to suggest the subject. Very early on, Chomsky and his students came to the conclusion that natural languages could *not* be generated by phrase structure grammars, or anything much like them. This Chomsky was not able to demonstrate, but his informal arguments appeared plausible. The grammars that he next turned to Chomsky termed *transformational.* In these, whole phrase structure descriptions – so-called tree diagrams – find themselves mapped to other tree diagrams, as when an active sentence is transformed to a passive sentence. Much effort was invested in the analysis of such grammars; the more they were studied, the less useful they came to seem. In recent years, Chomsky has modified and adjusted his concept of what counts as a grammar. Only one transformational rule – *Move!* – remains in place. The phrase structure rules have been banished. Chomsky talks now of grammars in terms of *principles* and *parameters.* His work has become dense, clotted with technical devices. Philosophers have traditionally regarded linguistics as a gentleman's pursuit, something like croquet. Now they know better.

But anyone can explore the details for himself. My interest is not with grammars at all.

Contra Force Majeure

Behaviorism flourished in psychology, and flourishes still in sociology and political science, in part, at least, because of the purely rhetorical skills of its practitioners. *Force Majeure.* In Chomsky, the behaviorists encountered a polemicist of comparable skill. *Contra Force Majeure.* In reviewing B. F. Skinner's *Verbal Behavior*, for example, Chomsky managed in one blow to empty Skinner's reputation of its brimming content. What gave Chomsky's review its terrific whump was its novel point of attack. The difficulty with the entire behaviorist analysis of language, Chomsky argued, was a matter of its general scientific inadequacy. Where the methods of behavioral psychology are precise, they are irrelevant. Where they are relevant, they are not precise. Behaviorists had long cultivated an image of themselves as a hard warrior host casting a cold iron eye on the excesses of the humanities. Now they were in the position to which they had so long relegated others. Someone was denouncing them – *them*! – for a kind of intellectual softheadedness.

Chomsky continued his attack on behavioral psychology (and everything else) in a fantastic, wide-ranging guerrilla action. There were journal articles and public lectures, newspaper interviews and television appearances. He was everywhere: in front of Skinner, who stood dumbfounded as Chomsky abused him directly and suggested that he was a simpleton; to his side, as in his devastating criticism of behavioral techniques in linguistics, which, he wrote, may be applied to natural languages "only in the most gross and superficial way;" to the rear, where he savaged behaviorism from the footnotes – "As to Dixon's objection to traditional grammars, since he offers neither any alternative nor any argument . . . there is nothing further to discuss."

Poor Dixon.

No one was used to this sort of stuff. It was one

thing to have the analytic philosophers insinuate that some logical scruple in the behaviorist program had yet gone unresolved. These people could find fault with anything. It was another thing entirely for the behaviorists to be told that their approach to language and learning was at once fatuous and forlorn.

The behaviorist position, Chomsky wrote in a footnote, – and after this I really will stop – "is not an arguable matter. It is simply an expression of lack of interest in theory and explanation." This is the case, he goes on to say, "in Twaddell's critique of Sapir's mentalistic phonology."

My only question: Who on earth is this Twaddell?

Nous Sommes Si Beaux

That winter a café on the *Saint Germain des Prés* was very popular. Morris Salkoff and I used to meet there late in the evening and sit on the glassed-in deck. There were always a good many pretty, slightly older women sitting together in pairs, looking very elegant and cool and distant, their gold bracelets stacked in rows on their thin forearms, and several plump, carefully groomed Frenchmen, wearing fuzzy cashmere sweaters underneath their jackets, reading the unreadable *Le Monde* with great seriousness or one of those old-fashioned French paperbacks whose edges required cutting. Down the street, *Les Deux Magots* was empty, the waiters standing glumly by the door, towels over their arms; here there was warmth, and the smell of beer, and the distant hiss of the espresso machine.

I asked our porcine waiter, who had a lustrous brown mole on his chin from which a short, stubby black hair grew in defiant isolation, why they were so busy.

"Nous sommes si beaux," he said gravely – we waiters are so handsome.

I had met Morris Salkoff some years before when I had stopped in Paris on my way to Trieste. Now that I lived there, we had become friends. He was an American from Brooklyn. It was plainly and forever a struggle for him to sit still. His fingers had a complicated, busy, independent life of their own; they were always in motion, tearing at sugar envelopes, playing with spoons, wandering over his face, or tapping out a fugue on the table top. His feet, too, were a part of this vigorous secondary system. When he sat, they jiggled robustly. Morris spoke English in a high-pitched slangy rapid-fire discharge, the sentences, when he was excited (always), heaped on top of one another. His French was letter-perfect, exquisite, really, except for the touch of an accent, and Frenchmen regularly stood back in amazement when he spoke to them, regarding his abilities as something of a Lourdes-like miracle, and would have believed, had the subject come up, that having mastered French so well, he could also raise the dead. He was tall and stooped, and he dressed for comfort, in clothes that fit irregularly and that he purchased in the flea market. In routine financial dealings he was parsimonious to the point of pathology, counting tips out in centimes with the pained argumentative air of a man surrendering his frostbitten fingers for amputation, and eating deliberately in the cheapest of restaurants, where couscous and other Algerian horrors were served. To his friends he was generous to a fault. On several occasions that year I would ask him to trudge out in the cold late at night simply to loan me a few hundred francs. He always did.

He had lived alone in Paris for many years, and while he was not a European, he was no longer an American. "I'm just a rootless cosmopolitan," he said, "a cosmobopolite," he added, babbling sadly.

Through some quirk of circumstances, Salkoff found himself in Paris with a good job as a linguist at the CNRS (Centre Nationale de la Recherche Scientifique). His work was not in Chomsky's line. Over and over, on those winter evenings, he would insist that Chomsky was a philosopher, and not a scientist.

"What do you mean?"

"There's no *data* there," he said earnestly, "it's all speculation."

His voice now rose as he gesticulated wildly with his long-fingered hands.

"They don't look at the *ev*idence. All they do is formulate *rules*, and conditions on rules. It's not science. Chomsky's a philosopher, not a scientist."

"Morris," I said, "what evidence *is* there really in linguistics? What do you look for? There are an infinite number of sentences in a natural language. You can't examine them all."

"You look at *words*. You go to the *dic*tionary. It never occurs to these guys to look at a dictionary. How can you have a science without *data*? It can't be done. You don't *start* with the theories. Geologists spent years and years collecting data. They went to the hills and collected rocks. When they knew what they were talking about, *then* they formed theories."

"But what about language acquisition?" I said.

"Who knows about that," said Salkoff explosively. "Who knows? We don't know. It's a mystery. But no one knew anything about plate tectonics until the data was collected."

I never knew quite what to say at this point; we talked of other things.

Often in the late afternoon, I would visit Salkoff in his office. He worked at the University of Paris at Jussieu, where I taught mathematics. The university itself consisted of nine high-rise towers that rose in terrible ugliness from a huge, permanently muddy slab of concrete – this in one of loveliest *arrondissements* (the fifth). Bordering the square roof of each of the towers was a doubled row of barbed wire. Although I had an office on campus,

I found the university so oppressive that I made it my business to be there as little as possible. Once I remarked to a friend that the place actually had the creepy feel of a prison or fortress. I discovered subsequently that the architect had been a survivor of a concentration camp, a man for whom the universe had blackened.

Morris kept his office on the ninth floor of the largest and bleakest of the university's towers. The lobby of the building was, of course, covered with misspelled graffiti of the corniest sort. Inevitably, some lost pigeon would be flying frantically from one end of the corridor to the other. The building's three elevators worked erratically, with a fine French sense of their own independence. By the time one would reach the lobby, a great crowd of irritable students would have gathered by the steel doors. When they finally opened, those doors, it was every man for himself in the crush. More often than not, when the elevator doors closed, the elevator itself would simply sit there, refusing to move, the doors remaining shut, the air becoming close.

After a long dispiriting trudge, I reached Salkoff's office – a little, overheated cubicle, with a view of the Pantheon and all of Paris beyond, gray on a gray day. In the drawers of his desk, Salkoff kept an emergency supply of rations. There were dried fruits, nuts, and a package of buttery cookies from Normandy to which I was especially partial.

"Take, take," Morris said, pushing the package toward me across the desk.

From the window I could see the lovely, crumbling, old houses of the fifth *arrondissement*.

After we had eaten, Salkoff energetically rummaged through a pile of papers on his desk until he located a neatly drawn graph.

"I need your help," he said imperiously, lifting a thin-leaded pencil.

At the time, I must mention, Salkoff was very busy

studying certain grammatical properties of the English verbs. No doubt he was the only man in Paris paid diligently to investigate a variety of English and American dictionaries. In his attitude toward syntax, Salkoff was very much under the influence of Maurice Gross, the director of the linguistics laboratory. Some years before, Gross, at the suggestion of M. P. Schutzenberger (these academic connections are complicated as kinship rituals in the tropics), had sent me a copy of his own paper on theoretical linguistics. Did I have any comments? Not many, in fact, and those that I had were all of the delicious in-English-we-do-not-say variety. But I found his arguments interesting, even striking. In organizing the fifteen or so thousand French verbs, he had discovered that with respect to a sample of obvious grammatical rules, each verb constituted an exception to some rule. Those grammatical rules that Chomsky advocated were designed to subsume any number of distinct cases by means of a collection of productive principles. How very embarrassing, then, to see the evidence so wayward, almost as if French grammar had been conceived as a rebuke to the scientific spirit. Whatever the ultimate role of theory in linguistics, Gross championed the primacy of the data – this on the grounds that, *au fond*, the facts come first.

Salkoff, I knew, had conceived an interest in the English verbs of motion. One of the peculiarities of his research was simply that in difficult cases he never quite knew whether a given construction was grammatical or acceptable. It was obviously no use asking his French colleagues. "They think that in English *anything* is grammatical." Indeed, a great number of otherwise cultivated Frenchmen imagine that the English language is somehow formless, and may be acquired by osmosis.

"What do you think of this sentence?" Salkoff asked. *"He elbowed his way through the crowd."*

It sounded fine to me.

"He shouldered his way through the crowd."

Still fine.

"He armed his way through the crowd."

"I guess."

"What about *He walked his way through the crowd?"*

This sounded definitely off. My inclination was to say that whoever He was, all that He did was to walk through the crowd directly, "His way" constituting a pleonasm.

How very unusual, Salkoff observed tartly, that only certain English verbs accept "his way" as part of "He X'd (moved, shouldered, walked, butted, elbowed, etc.) his way through the Y" construction. Very quickly, I would reach a point of troubling insecurity in my own intuitions. In general, Salkoff thought, we tend to reject as ungrammatical rather too many linguistic constructions. When I was quite sure that some or another expression was simply not English at all, Salkoff would search his files and discover that the thing had appeared in *The New York Times*.

Two Great Frauds

Richard Montague was a small, very dapper, compact, cufflink of a character. He was dressed in a neat blue suit, a snowy white shirt, and a matching crimson tie. We had met for drinks in mid-town Manhattan – he, Daniel Gallin, and I. His hands, I noticed, were square, the fingernails manicured and covered with a clear polish. A logician by profession, Montague had a reputation for great technical brilliance. His papers were adroit, carefully written, biting, and completely beyond the intellectual grasp of all but a handful of analytic philosophers.

For some reason he was ill at ease that afternoon, and looked fitfully around the hotel's bar, as if he suspected somehow that nothing was going to turn out properly. Beyond the bar, in the lobby of the hotel, there was an absurd canary cage in which a pair of yellowish birds were cheeping nervously, complaining, I am sure, about the price of drinks or room service.

We talked of taxes and politics and How on Earth do you survive in this place – meaning New York. Then the discussion turned to mathematics and Montague cheered up. He had just commenced his research program into formal grammars and had published a series of papers of truly monstrous technicality. He liked to imagine that he and Chomsky were rivals.

"There are," he said, "two great frauds in the history of twentieth-century science. One of them is Chomsky."

I reached for the peanuts.

"And the other?"

"Albert Einstein," Montague said decisively, glad that I had asked.

PART III

Formal Systems

The world, a Japanese poet once remarked (I am relying here entirely on James Clavell's *Shōgun*), is but a dream within a dream, a circumstance that to my mind suggests a nightmare more than anything else – imagine clambering arduously from one dream only to emerge within another. The image of a dream within a dream suggests a certain ontological profligacy, with as many worlds popping up for inspection as there are dreams, another reason never to fall asleep. When it comes to worlds, my own inclination is to affirm that there is only one; but the idea of a motley is suggested, in fact, by much of modern science. In General Relativity, space and time are curved like a worm (a fairly *large* worm, of course); there are Black Holes and, as in a fraternity party, things begin (and end) with a bang. In quantum mechanics, on one interpretation at least, whatever is possible is actual, so that in some hidden, humble dimension of space or time, Benedict Arnold finds himself restored to perfect respectability, the unpleasantness about those secret codes forgiven and forgotten. How the two disciplines interact, no one knows. In a world somewhere between General Relativity and quantum mechanics, the universe emerges as a thing of strings – dense little coils sunk into six hidden dimensions. And in still other worlds, *symbols* – words, numerals, shapes – are manipulated according to fixed, precise, literal, and explicit rules. Such are the *formal systems*. Here, as in dreams (to revert to the same metaphor), space and time count for little. The formal systems are *abstract*, and share a queer kinship with other abstract objects – games such as chess, checkers, or the inscrutable Go, works of the graphic arts, poems, novels, number systems. And although a formal system is man-made, like any work of fiction the thing is often capable of surprising its creator, as when a character in a novel turns to drink to the astonishment even of the author.

In speaking of the formal systems, the logician has in mind two vantage points – the Outside, a perspective that includes the genial symbolmaster, aware, as he must be, of the game and its rules, but also of the game and its goals; and the Inside, wherein the symbols dwell, and which includes no reference to an outside at all. From a rook's point of view, the universe exists in two dimensions; the piece inhabits a world made up of other pieces and the squares on which they move. Life on this level is a matter of the right move made at the right time, with the moves divided neatly into the licit and illicit. The linguist's grammar is so conceived. From the player's outside point of view, there is more, of course. The richly tessellated plane divided into contrasting squares of light and dark and the heavy, oiled, wooden pieces are means merely to express one's personality, or to subdue an opponent, or to express the ancient rivalry between the doubled and divided forces of the universe, or simply (in my own case) to pass the time.

Like a melody or motif, the very idea of a formal system suggests more than it implies. Under one interpretation, as I have said, ordinary games (chess, life) emerge as formal systems of a certain sort. With its hardware imaginatively wiped away, the digital computer collapses into its program, and programs, too, are formal systems. But beyond any of its instances, the idea of a formal system evokes as an echo a prior distinction between form and content; this is one of those fateful pairs of matched ideas that make an appearance at every stage in western culture and that can consequently be presented like well-known, well-worn antagonists: faith and reason; energy and matter; fields and particles. It is within logic that the distinction between form and content is drawn most scrupulously. Both mathematical and traditional logic treat of arguments. Those that are *valid* are accorded a special point of distinction. In a valid argument, true premises lead inexorably to true conclusions, the inexorability of the deduction arising as the result of a force with no known cognate in physics. Talk thus of truth suggests that validity hinges on meaning or interpretation. In fact, the validity of an argument does not depend *directly* on the truth of its premises: There are valid arguments in which false premises lead inexorably to false conclusions. The peculiar and overwhelming force generated by a syllogism arises from the *general* fact that if all *A*'s are *B*, and *x* is an *A*, then *x* must be a *B* as well. Here the specific content associated to the argument has dwindled to a tapering point and seems in danger of vanishing altogether. Yet some concept of content plainly persists in the presentation of even the most schematic argument. Those silent dummy letters (*A*, *B*, *x*) refer to something: The logician is simply indifferent to their interpretation. And although the validity of an argument and the truth of its premises are slightly separate matters, validity is ultimately defined in terms of a conditional concept of truth; and truth is a concept with plenty of content of its own.

It is this last lingering almost aromatic commitment to some form of content that vanishes in a formal system: In the end, the thing, like a skyscraper by Mies van der Rohe, appears to achieve a wholly Platonic and thus wholly pathological existence. Here numbers exist as numerals, words as letters, and letters as specified shapes. Imagine, for example, the world contracted to two letterlike shapes, *S* and *a*. These symbols the logician inscribes on the lines of some imaginary sheet of paper. The behavior of the symbols are governed by just four rules:

1) *S* may be written on the first line as an initial symbol;

2) $S \rightarrow a;$
3) $S \rightarrow a + S;$
4) The system is to stop after any line that does not contain S.

The arrow ($\rightarrow$) indicates that the symbol S may be rewritten as the symbol a whenever S occurs, or, to introduce an element of choice into the system, as the symbols $a + S$.

This system comes to vivid life when the rules are allowed to play over the alphabet in a derivation (geographically I am now traversing that country in which the grammars form a state). The numbers at the right of the derivation indicate which rule governs which step. They are not part of the derivation itself:

1) S	Begin	(Rule 1)
2) $a + S$	Continue	(Rule 3)
3) $a + a + S$	Continue	(Rule 3)
4) $a + a + a$	Continue	(Rule 2)
5) $a + a + a$	STOP	(Rule 4)

The derivation, of course, need not to have come to a halt at 4). This little system is capable of generating an infinite number of distinct objects: All sequences of the form a, $a + a$, $a + a + a$, . . . , in fact.

•••

At Princeton, graduate students in philosophy were required to know at least some logic for their examinations. Those who wanted to get by with the least work studied Quine's little book on the methods of logic. The material could be learned in a lazy afternoon and, once mastered, afforded a student the agreeable impression that he had acquired a valuable argumentative skill. Students with a desire for punishment attended a course in mathematical logic offered by the department of mathematics. This was taught by the great American logician, Alonzo Church.

I showed up that first day not knowing quite what to expect. Presently Church entered the lecture room. He was then a man in late middle age, tremendously stout, his great stone face set solidly on a tree-trunk neck and a Sherman-tank torso. His hair was cut short in a spikey crew cut that gave to his features a look of vague lunacy.

He introduced himself. This was, he said, to be a course in mathematical logic, or, to be more precise, formal logic from a mathematical point of view. He presumed that while we might know nothing about logic, we all had a certain level of mathematical sophistication. The mathematicians in the room now bobbed their heads in unison. Some of the material would be quite difficult, but, while it required talent to *establish* a theorem in mathematics, any idiot – he paused to survey his students – could with diligence come to *understand* the result. He would be lecturing from his own textbook. He preferred that we not ask questions. It would be wise if we copied every word. He began to write.

That day, and all the days that followed, Church taught by transcribing the theorems, proofs and definitions from *An Introduction to Mathematical Logic* onto the blackboard. As he wrote, he read aloud what he had written, often pausing to refer to the very footnotes in his book. I have never observed a man so curiously self-absorbed and so self-satisfied. The process of lecturing Church plainly regarded as a form of self-expression – this understood quite literally. What Church said, he said to himself alone, often pausing to revise what he had written on the blackboard so that it conformed better to his inner vision of what he needed to say. These slow-moving emendations, of course, made note-taking a misery, for no sooner had we diligently copied a blackboard full of symbols, than Church would pause, reflect, and then, with surprising energy, erase the whole jungle, quite without explanation, in order to begin again from scratch, the only difference in the presentation a matter of the re-arrangement of the color code he used to indicate typefaces. It would be a mistake to imagine that these didactic methods were ineffective. The great glacial slowness by which instruction proceeded was soothing; Church's deliberate manner, while it conveyed nothing of the excitement of mathematics, managed to suggest in its operation an intelligence that in power made up what it lacked in speed.

During that first semester, Church discussed a symbolic system – the *propositional calculus* – in which ordinary sentences are treated as indecomposable units and symbolized by means of sentential or propositional variables: P, Q, R, S, for example. It is the purpose of this calculus to codify the various inferences that depend only on sentential structure. Any sentence, speaking now of ordinary language, implies itself. The machinery of this inference, and all others like it, the propositional calculus must make clear. Now from the logician's perspective, the propositional symbols are indecomposable: They have no internal structure and are understood as standing indifferently for sentences distinguished only in point of truth or falsity. This suggests an elegant economy: Rather than standing for sentences, the sentential symbols might as well stand for truth values – T or F. And to the extent that truth and falsity are arranged in pairs, *any* pair of objects would do as well – 0 and 1, say.

In mathematics it is usual to mark a distinction between the axioms of a given system and the system's theorems. It is the axioms that must be taken on faith, at least with respect to the possibilities for their proof. The theorems are what follow logically, and so must be true if the axioms are true. The mathematical distinction between an axiom and a theorem corresponds to the ordinary division of statements into those that are assumed and those that can be proved. In a formal system, this distinction is robbed of its purely mathematical content: An axiom within a formal system has a purely *geographical* significance: It marks the spot where the system starts. Correspondingly, the theorems mark the spot at which the system stops. No concept of meaning comes to infect either the axioms or the theorems. In investigating a formal system, the mathematician, Church explained, aims simply to study the process of moving from one statement to another – inference on the wing.

Yet in urging upon us his stern renunciation of meaning, Church also demanded that we keep the intended interpretation of the symbols firmly in mind, and thus promptly engaged the class in a delicate form of doublethink. Certain statements of the calculus, for example, turn out to be true regardless of the truth of their sentential constituents: It is *always* the case that if *P* then *P*, whether or not *P* is itself true. These are the *tautologies*. Now the notion of a tautology is purely semantical: In explaining the idea, I go beyond the symbols to make contact with their reference. The theorems of a formal system, on the other hand, are defined without appeal to meaning. If the propositional calculus is to have any point or purpose, the theorems that it sanctions must coincide with its tautologies. They do. The system is *complete*. But the manipulation of a formal system involves a sly maneuver (akin, actually, to the practice of certain forms of religious orthodoxy) in which the logician simultaneously accepts and transcends the system's limitations.

After a number of weeks we had managed in class only to prove certain propositions that were in their meaning more obvious than any of the axioms. When reservations were raise, Church said cryptically that what could be expressed *within* a formal system was less important than what could be said *about* a formal system and left the matter at that.

Only years later did I come to understand this remark.

The Hilbert Program

A difficult section is coming. More than most men, mathematicians and artists alike often attempt to catch themselves in the act of creation, like patients in certain schools of psychoanalysis who endeavor to witness their own conception. Mathematical structures that shed an inward light on the very activity of mathematics have inevitably exerted an unwholesome fascination on the mathematical community. In a well-known remark, David Hilbert referred to the theory created by the melancholic Georg Cantor as a "paradise." What prompted Hilbert's enthusiasm was Cantor's transcendental hierarchy: Sets, sets of sets, sets of sets again. All of elementary arithmetic may be reflected from these cascading classes.

Gottlob Frege argued for something even stronger than the reflection of arithmetic from within set theory. What he envisioned was a kind of fantastic reduction of arithmetic to a form of logic. The reduction complete, Frege hopefully affirmed, only the axioms of mathematical logic would remain, unassailable as bedrock. Unhappily, Frege thought of set theory as a part of logic, and set theory proved inconsistent early on, with trouble arising at the schema "x belongs to x," precisely the spot at which Cantor's theory endeavored to swallow its own tail.

Beyond this, it has become plain in recent years that the axioms of set theory exhibit a disturbing waywardness. The continuum hypothesis is independent of the other axioms of set theory; so, too, is the axiom of choice. This makes for a multitude of set theories, each with its own sparkling academic credentials. Yet when it comes to sets, mathematicians believe that there is but *one* axiomatic collection of truths about them. They are thus in the position of a man who worships one God but pays his respect to at least two Popes.

In high-school geometry a distinction is drawn between the axioms and the theorems of a geometric system. The theorems are demonstrable; the axioms have to be accepted on faith. Some students see only an unhappiness in this epistemological division. If anything is to be accepted without proof – so goes their argument – why not the theorems themselves? Still, many geometric statements are anything but obvious. Deriving them from the axioms satisfies some intellectual hunger; the confinement of assumptions to a single set offers the mathematician the illusion that his scientific imponderables have been minimized.

The axiomatization of geometry is an achievement of Greek thought and for many centuries remained the object of intense (and justified) scholarly veneration. The discovery in the nineteenth-century that this great edifice was marred by a number of axiomatic cracks came as a troubling surprise, and suggested somehow that, when scrutinized closely, absolutely *anything* might reveal its logical infirmities. The German mathematician David Hilbert constructively examined the foundations of Euclidean geometry during the 1890s. Those logical cracks that he had discovered he covered carefully. Hilbert effected his repairs at a time of revived interest in axiomatics generally. On the Continent, the Italian mathematician Giuseppe Peano had turned his limpid gaze toward the natural numbers: the set 1, 2, 3, . . . and all points beyond. If anything, the natural numbers lie even closer to intuition than the shapes and surfaces studied in Euclidean geometry. In providing a plausible set of axioms for arithmetic, Peano and his students seemed to be moving toward a formal explanation of an ancient aspect of the human mind.

Peano expressed his axioms in arithmetic terms: The number 0 appears there, and so does the notion of a successor to a natural number. To some theorists, this seemed insufficiently foundational. In *Principia Mathematicia,* Peano's axioms disappear in favor of assumptions that are severely logical, with logic construed so as to include set theory and the theory of types. Frege's work yet lay unburied; the theory of types, Russell and Whitehead introduced to excorcise the antinomies of set theory. The axioms of the *Principia* must still be accepted on faith, of course – else why call them axioms? – but their ostensible character as principles of logic (rather than arithmetic) made them seem especially obvious. The three volumes of the *Principia* were published by Bertrand Russell and A. N. Whitehead in 1913. This may be taken as the last date that anything, anywhere seemed certain.

A set of axioms together with its logical consequences make up an *axiomatic* (but not necessarily a formal) system. Consistency is a concept that counts heavily in the evaluation of such systems. The appearance of a contradiction in an axiomatic system is a severe embarrassment: From inconsistent premises, anything follows, and hence anything goes. There are infinitely many theorems in Euclidean geometry; counting noses to make sure that no contradictions crop up is an enterprise with little appeal. Hilbert *did* manage to show that Euclidean geometry is consistent by interpreting geometrical relationships in arithmetic terms, but for all its ingenuity, Hilbert's proof is unsatisfying. It shows only that the consistency of geometry hinges on the consistency of arithmetic – cold comfort to anyone prepared to doubt the consistency of arithmetic.

A set of axioms, Hilbert observed, is nothing but a series of statements with an intended interpretation or meaning – in the case of arithmetic, an interpretation in the world of numbers. Their meaning the mathematician may artificially suppress. With meanings etiolated, an axiomatic system becomes a formal system merely, a kind of game. But the concept of a proof, and hence the notion of a consistent system, may yet be given sense of sorts entirely within a calculus of shapes. With axioms understood as syntactic shells, a proof of a particular proposition emerges as nothing more than a specific sequence of axiomatic transformations. (It is this maneuver that serves to draw a connection between mathematical and *mechanical* ideas.) Now a consistent system is one in which no proof of a proposition and its negation occurs. In a formal system, propositions are defined by their shape, and so is the sign for negation.

This suggests a tactic by which the consistency of arithmetic might be established *absolutely*. If an axiomatic system is inconsistent, all statements that may be expressed within the system follow profligately from the axioms – this by elementary logic. To establish the consistency of an axiomatic system it suffices that a particular proposition – $0 \neq 0$, to take an arithmetic example – *cannot* be demonstrated as a theorem. A *metamathematical* proof of the consistency of arithmetic sticks to syntax and establishes that no symbol of a certain shape appears as a theorem if its negation does so as well. But plainly what holds for a system of shapes holds for the same system understood as a series of signs denoting real mathematical objects. Contradictions require symbols to express them. If no such symbols exist, neither can the contradictions.

Arguments that involve stripping a system to its syntactic shell Hilbert called *finitistic* – not entirely a happy term, but one that suggests what he had in mind. The canny mathematician needs somehow to avoid the regress of relative proofs if he is to make a persuasive case for the consistency of arithmetic. But a persuasive case must also avoid metamathematical methods of proof as susceptible to inconsistency as those of the original axiomatic system. The symbols of an axiomatic system, taken solely as shapes, do have a kind of palpability – they lie there on the page. Arguments that trade on their indubitable *presence* seem relatively sturdy, especially when measured against the far more abstract arguments that figure in arithmetic itself.

Brunch

Brunch was to be served. We climbed the steps of the narrow frame house and rang the bell, which sounded somewhere in an intolerant cheep.

Janet Gallin greeted us at the door. She was lanky and elegant, with a thin face and a fabulously narrow, exotic nose – the thing had a striking, marvelously attractive life of its own – and painted, pointed fingernails.

"Well *hello*," she said sunnily in a diphthonged drawl, "You must be the Berlinskis." Turning especially to me, and then showing me the sides of her very shapely ears, one hand holding her hair back, she chirped, "You look absolutely *mis*erable. Don't you just *hate* meeting people?"

With no time allotted for my immediate assent, she draped one elegant hand over my forearm, and propelled me into the living room.

"I want you to meet some people," she said firmly.

We were introduced to an assortment of couples and passed in short, choppy stages through the inevitable brunchy ritual: Bend down, shake hands, a tentative bird-fluttery Hi, a few mellow Hellos and a Glad to See You, here and there a bone crusher of a handshake or one that was moist and lingering, a real-estate salesman and a physician, and a woman in a red tafetta dress, and a ten-year-old child with straight black hair and glowing eyes, and a woman who wanted to know whether I had been to Israel, and a man (an editor, I think) who was telling a small group how he had once traveled from San Francisco to Los Angeles in the company of Tina Turner.

"Honey," said our hostess, to whom I had formed an instant attachment, "these are the Berlinskis."

The Berlinskis turned, like the obedient tag team they were, to meet and greet Daniel Gallin, Janet's husband, who had, I now noticed, been listlessly circling her sun like some pained outer planet until fetched from orbit by that resonant "Honey."

He was tall, with a carefully shaved moody face that was longer than it was wide, dark shoe-polish eyes, sensuous, somewhat extravagant lips, and another noble nose. They grew them like roses in that family. As he stood before us looking down, I noticed his wonderfully expressive hands, the fingers long and tapered.

"Let's get away from these dumb shits," he said.

The Connoisseur of Complexity

An axiomatic system is *consistent* if it admits of no contradictions; *complete*, if proofs can be produced for any proposition the system is capable of expressing. An incomplete system is one in which an *undecidable* proposition figures. Such propositions outrun the axioms in a disagreeable way. A proposition that is undecidable must be true if its negation is false, or vice versa. In either case, some truths about a mathematical subject cannot be derived from the axioms. But the axiomatic method was itself justified via the assumption that the axioms themselves constitute the *only* class of indemonstrable truths.

Whitehead and Russell provided an axiomatic version of arithmetic in *Principia Mathematica*. Hilbert asked for metamathematical proofs that such systems were both consistent and complete. Many mathematicians devoted themselves to Hilbert's program in the years from 1918 to 1931. Some sense of optimism still shines through their research papers; the succession of partial results – by

Ackerman, von Neumann, Herbrand, and others – suggest in retrospect the confident activities of scholars convinced at last that they were getting the foundations of arithmetic in good working order.

In 1931, the Austrian mathematician Kurt Gödel published his monograph "On Formally Undecidable Propositions of *Principia Mathematica* and Related Systems" in the German scientific periodical *Monatsheft für Mathematik und Physik*. It ran to only twenty-five pages and appeared incomprehensible, even to mature mathematicians. To those capable of understanding them, the conclusions of this paper had the force of cannon shot. The Hilbert Program, it turned out, was an impossibility. In any formal system rich enough to express the propositions of arithmetic, there exist sentences that can neither be proven nor disproven, and that simply hang there in a state of petulant indecisiveness. Adding them to the axioms does no good. Arithmetic is incomplete and *incompletable*. Some logicians could not bring themselves to believe the plain evidence of their senses, and kept poking hopefully at the corpse of the Hilbert Program. Some are poking at it still. Elsewhere, Gödel's great theorem is accepted for what it is – a cold wind over dark water.

•••

An axiomatic system *S* is *adequate for arithmetic* if *S* has the power of Peano's axioms, whatever the details of its formulation. For any such system *S*, Gödel established that a proposition *G* of *S* exists such that neither *G* nor its negation is demonstrable in *S*. *G*, while it cannot be demonstrated *within* *S*, is nonetheless true, and true, moreover, of just the objects that *S* itself describes.

The statements of *S*, like statements in English, are built from a primitive stock of symbols, which correspond roughly to ordinary English words. Since the subject of Gödel's proof is syntax, some special scrupulousness is needed in listing them. For my purpose, I need only assume that such a listing exists, and that, moreover, those sequences of primitive symbols that make syntactic sense have been segregated from all the rest. These are the *well-formed formulas*. They correspond to the grammatical sentences of English.

Understood in its ordinary sense, *S* is an axiomatic system for the expression of arithmetic truths; understood as a formal system, *S* is a collection of symbols that have been shorn of their meaning. In talking *about S*, the logician has both perspectives in mind. What he has to say – the logician, I mean – is said in a meta-language, one whose subject is *S* itself. To the extent that his concerns are purely formal, the logician confines himself to the system's syntax. This is the perspecitive that the Hilbert Program requires. Yet a magical, mysterious, Talmudic connection exists between what the system *S* says (arithmetic) and what the logician means to say (metamathematics). When the statements of *S* are given their ordinary arithmetic interpretation, they can be made to represent their own syntax. The distinction between form and content, upon which logic itself rests, now begins to waver or wobble.

The primitive vocabulary of *S* includes arithmetic constants or names – symbols that pick out a fixed number – and arithmetic variables – symbols whose reference is indeterminate, in the fashion of 'x' in the algebraic equation '$x + x = 2x$'. There are also symbols for arithmetic relations and properties, marks of punctuation, and so on. In general, 'x', 'y', 'z', are variable; 'm', 'n', and other mid-alphabetic letters, constant.

Suppose that to each of the primitive symbols a number is assigned. This is its *Gödel number*. The assignment satisfies the condition that every well-formed formula has exactly one Gödel number; different well-formed formulas have different Gödel numbers. The numbering applies not only to specific formulas of *S* but to sequences of formulas as well. A proof of a proposition within S is nothing but a sequence of formulas. Proofs receive their own Gödel number.

The details by which Gödel numbers are actually assigned are uninteresting and complex. The chief point to the assignment can be set out without a full description of the method. I annunciate it as a principle to give it an appropriate aura of importance. By the *translate* of a metamathematical statement *about* S, I mean a purely arithmetic statement *within* S. The principle that I promised is in two parts:

I Every metamathematical statement about S has a translate within S;
II If P is a metamathematical statement about S, and P^* is its translate within S, then P is true if and only if P^* is true.

Consider now the metamathematical assertion that

1) The sequence of formulas H is a proof of the formula F.

Both H and F are syntactic items: Both are assigned Gödel numbers, m and n, say. 1) can thus be rewritten as

2) The sequence of formulas whose Gödel number is m is a proof of the formula whose Gödel number is n.

2) is a metamathematical statement still: It identifies H and F in a roundabout way by means of their Gödel numbers, not their names. In virtue of I, 2) can itself be represented *within* S by means of an arithmetic statement. The metamathematical relation "is a proof of" figures in both 1) and 2). Say that its arithmetic counterpart is dubbed DEMONSTRATION. 2) then finds its arithmetic voice in

3) m DEMONSTRATES n,

or

4) DEM(m,n),

where 3) and 4) are statements similar in form to

5) 7 is greater than 5.

In fact, 3) and 4) share more than a similarity in form to 5): They are both statements *about* some arithmetical relationship. But the Gödel mapping insures that 3) and 4) speak to syntax. Thus 1) and 2) are true if and only if 3) and 4) are true, and vice versa.

The simple device of assigning numbers to formulas has the result that statements about *formulas* can now be expressed by statements about *numbers*, another matter entirely, but one strangely coordinate to syntax.

At 1), the concept of provability was expressed by metamathematical means; at 4), in purely an arithmetic way. 4) refers to particular numbers – m and n, in fact – but there is no reason to be so specific. In

6) DEM(x,y),

the variables 'x' and 'y' stand in for the constants 'm' and 'n'. But suppose, now, that *no* number stands to y as m stands to n in 4). In arithmetic terms:

7) There is no number x such that x DEMONSTRATES y

or

8) (x) ~ DEM(x,y),

where '(x)' denotes universal quantification – the logician's "for every x" – and '~' denotes negation. 8) is again a statement of arithmetic; it has an obvious metamathematical counterpart:

9) There exists no proof of the formula whose Gödel number is y.

Consider, now, a particular formula of *S*, *F*, say, and suppose that within *F* the variable '*y*' occurs. *F* itself has a Gödel number, and so does '*y*'. Assume that the Gödel number of *F* is *m*, and that the Gödel number of '*y*' is 17. These facts I now place on hold.

In the manipulation of a formal system, the logician often needs to effect an alteration within various formulas. Having demonstrated, for example, that *if P then P*, the scrupulous logician risks standing baffled before *if Q then Q*; even though the formulas are identical, the variables are different. A proof of the first formula, speaking strictly, does not yet count as a proof of the second. To make good this obvious deficiency, the logician countenances various rules of *substitution*. These are defined – in the meta-language, of course – so that well-formed formulas remain well-formed under substitution.

In particular – I have lifted the hold; your operator is now back – the formula *F**, which results from substituting the numeral '*m*' for the variable '*y*' in *F*, is well-formed. Thus *F** is a formula of arithmetic and as such has a specific Gödel number – *m**, say. *F** may also be described metamathematically by means of its Gödel number:

> 10) The formula whose Gödel number is *m** is derived from the formula whose Gödel number is *m*, by substituting the numeral for *m* for the variable whose Gödel number is 17.

I and II establish that 10) may be expressed entirely within arithmetic. 10) mentions a complicated operation of substitution. That metamathematical operation is again reflected within arithmetic by purely an arithmetic function – SUB, say – of the numbers *m* and 17. Given *m* and 17, SUB acts to name the Gödel number of *F**, which, of course, is *m**. Thus

11) $m^* = \text{SUB}(m, 17)$,

which is a statement about the natural numbers, saying, in straightforward fashion, that two of them are identical. But under the conventions of the Gödel mapping, 11) also identifies m^* as just the number of the formula that satisfies 10).

Nor is there any real need to make 11) quite so specific: 'm^*' and 'm' may themselves be replaced by variables, as in:

12) $x = \text{SUB}(y, 17)$,

which speaks of a Gödel number x but identifies x only as the number of that formula which results from substituting in the formula whose Gödel number is y the numeral naming y for the variable whose own Gödel number is 17. A particular number is designated by 12) only when y is specified, as in 11).

The formula displayed at 8) says merely that no proof exists for the formula whose Gödel number is y. Not knowing what y is, we cannot know which formula remains insusceptible to proof. This indeterminacy results from the fact that 'y' is variable; but then again, so is '$\text{SUB}(y, 17)$', since 'y' figures in this formula as well. By an interchange of variables,

13) $(x) \sim \text{DEM}(x, \text{SUB}(y, 17))$,

which says that

14) No proof exists of the formula whose Gödel number is $\text{SUB}(y, 17)$.

The difference between 13) and 14) is just that the first is an arithmetic and the second a metamathematical statement.

As a statement of arithmetic, 13) has a Gödel number, m suppose. The indeterminacy that infects 12) infects 13) as well, since there is no way of telling from the formula which proposition cannot be demonstrated. Substituting a constant for 'y' in 13) fixes the formula on a particular number. There is no reason why 'm' itself cannot replace 'y', as in the following formula, G:

15) $(x) \sim \text{DEM}\ (x, \text{SUB}(m, 17))$.

It is G that asserts that no proof exists of the formula whose Gödel number is SUB(m, 17). As a formula of arithmetic, G is provided with its own Gödel number, g^*, say. Astonishingly enough,

16) $g^* = \text{SUB}(m, 17)$!

the exclamation point at the end of 16 serving to take astonishment to a higher power.

The argument for the identity expressed by 16) involves a calculation along the equation's right and left sides. Thus

17) SUB(m, 17) is the Gödel number of the formula that results from substituting the numeral naming m for the variable whose Gödel number is 17 in the formula whose Gödel number is m.

But

18) g^* is the Gödel number of the formula that results from substituting the numeral naming m for the variable whose Gödel number is 17 in the formula whose Gödel number is m.

– this because 15) was obtained from 13) by substituting 'm' for 'y' in a formula whose Gödel number is m.

G is a formula of arithmetic. It has thus a metamathematical meaning. Its form puts one in mind of 13), and so G can be interpreted as saying that no proof exists of a formula with a certain Gödel number. But in view of 16), that Gödel number is precisely the Gödel number associated to G. Thus G says of *itself* that

19) G is not demonstrable.

In the meta-language:

20) The formula G whose Gödel number is SUB(m, 17) is not demonstrable.

There is a strong suggestion of paradox here; but, in the end, no paradox at all. Gödel showed that if the axioms of arithmetic are consistent, then G remains unprovable. Actually, what Gödel demonstrated was that if there exists a proof of G in S, then there exists a proof, too, of its negation. This follows almost at once from the peculiar self-referential properties of G. For suppose there were a proof of G in S. Then there would be some formula, k, say, such that

21) DEM(k, SUB(m, 17)),

since 'SUB(m, 17)' names the Gödel number of G. By elementary logic, this implies that

22) $\sim(x) \sim$ DEM(k, SUB(m, 17)),

which is the denial of G. Thus

23) If G is demonstrable, then so is $\sim G$.

This would mean that

24) S is inconsistent,

so reasoning backward, by contraposition,

25) G is unprovable.

In virtue of what G says of itself, namely that it is unprovable, 25) shows that G is true. But G is an ordinary statement of arithmetic whose powers of self reference emerge under the Gödel mapping. Taken in its usual sense, it refers to some complex but nonetheless quite tangible arithmetic power or property. G is thus a true statement of arithmetic obdurately out of the reach of the axioms.

Finally – and this is the second part of Gödel's theorem – the unprovability of *G* infects attempts to establish the consistency of *S*. In particular, 25) hinges on the assumption that

26) *S* is consistent.

This is a metamathematical claim expressible once again in arithmetic. *S* is consistent just in case there is at least one formula in *S* that cannot be demonstrated. Talk thus of indemonstrability recalls the form of the formula at 8). Using that formula as a guide makes for a way to represent 26). Say that the formula that results is *C*. Thus, to the extent that *C* expresses the consistency of arithmetic,

27) *C* implies that arithmetic is incomplete,

by virtue of the unprovability of *G*. But the consequent of 27) also admits of arithmetic transmogrification simply as *G* itself. Thus

28) *C* implies *G*.

Now as it happens 28) – the whole thing – is a demonstrable formula of *S*. But from this it follows that

29) *C* is not demonstrable,

since if it were, *G*, by modus ponens, would be demonstrable as well, and by 25), *G* is *not* demonstrable.

C is a formula, recall, that expresses the very consistency of arithmetic. The establishment of 29) shows that any proof of the consistency of arithmetic cannot be expressed entirely within arithmetic. Metamathematical proofs of the consistency of arithmetic involve forsaking finitistic arguments, in any sense of finitistic, however loose.

And with this it becomes plain that the Hilbert Program is an impossibility.

•••

Kurt Gödel died on January 14, 1978, in Princeton, New Jersey. In this century, it should be noted, Gödel's achievements in pure thought are virtually preeminent. Only Einstein's theory of general relativity represents an accomplishment of comparable intellectual grandeur. Of the actual course of Gödel's life I know little. Gödel took his degree in mathematics at the University of Vienna and remained in residence at the Institute for Advanced Study from 1938 until his retirement in 1976. By all accounts, he was a solitary and retiring figure. He taught no classes and offered rare public lectures. Those that he gave were accessible only to specialists. He held no office hours and did not concern himself with the rights of women or minority students. He was aloof and distant toward his colleagues. He failed to receive a full professorship at the Institute until 1953. A picture of Gödel in a volume published to commemorate his sixty-fifth birthday shows him standing against a blackboard. His face is utterly devoid of expression. His eyes are as bleak and pitiless at the sun.

Fillmore Street

In my hot youth I lived in San Francisco, a city that then (especially then) looked as if it had been rosily kissed, and that wore in public a dramatic tropical blush, visible most notably at twilight from my window, which overlooked the Golden Gate and the austere, tender-green hills of Marin.

I had not much to do. My doctoral dissertation, which I now regard with a moan of shame, I had written the year before; my classes at Stanford occupied only six hours of my time each week. Friends in medical school talked of deadlines and examinations and autopsies, and appeared at social events, when they appeared at all, with great, blackened circles underneath their eyes and a haggard expression. I took to running along the marinas in the cool evenings. I ate health foods. I translated some poetry from the German. I slept well at night and never regretted my decision to learn nothing of the law or medicine.

Under the impression that my intellect required more space than my apartment provided, I rented a small studio on Fillmore Street, and consecrated it to deep thoughts. The studio itself I furnished in a severely minimalist style. There was a table made of an old door, which rested on a pair of metal sawhorses, a very severe, straightbacked chair, and an enormous industrial carton, which served to collect trash.

Daniel Gallin and I would meet there every morning to work on mathematical logic.

I had known Daniel socially for only a few months, but we liked one another and sensed, I think, that our intellects were congeneric. For many years, Daniel had been engaged in an epic struggle to obtain a doctoral degree in mathematics from the University of California. This was an enterprise, he was convinced, in which the Almighty took a singularly perverse interest. Whole years disappeared as he searched the literature diligently for a suitable thesis topic. He could not prove what he wished, and did not wish to prove what he could. When he finally settled on an issue in modal logic, his bad luck held true to form. No sooner had he established a result, and boasted of his accomplishment quietly, than he would discover that someone else had actually established something stronger and more elegant and more to the point. His professors encouraged him warmly in his work, and kept at him during difficult times by offering him athletic-coach pep talks. Just as it appeared that Daniel might actually persevere, his thesis adviser died dramatically, leaving, Daniel was convinced, a Tut-like curse on his unfinished work.

In his personal life, Daniel was on the best of terms with any number of hypochondriacal complaints. When he was not suffereing from trigeminal neuralgia, or migraines, or mysterious and disabling orthopedic afflictions, or the symptoms of incipient coronary artery disease, or diverticulitis, he was much concerned with the laws of Kashrut, the meaning of his Jewish heritage, or some vague neurotic prompting that he was doomed – this last the perfect truth, unfortunately.

He was very funny, Daniel was, with a style of wit in which show-business references would be modulated by a fantastically controlled and perfectly pitched sense of elegance – the whole emerging in conversation (or on paper) as a bubbling cascade. He did not much care for life in California and insisted that he had been born in the wrong place: His goal was to tunnel backward in time toward eighteenth-century Lithuania or Poland or Latvia, where bearded Jews in caftans gathered companionably for morning prayers in muddy country villages, their voices rising into the smokey air, out of tune, reedy. It was difficult for me to reconcile this craving with his aversion to any form of physical discomfort.

I knew nothing of mathematics, really, except that it looked more difficult than analytic philosophy, but the problem in which we were both interested was plainly mathematical. Our aim was to investigate two slightly different definitions. In one – due to Alfred Tarski – the logical truths to a formal system are just those sentences true in every model. This is a definition in which the concept of a logical truth is tied to the mathematician's *world*. W. V. O. Quine, on the other hand, had argued that the logical truths were just those sentences that were true and remained true regardless of the syntactic substitutions made for their non-logical parts. This is a definition in which the concept of a logical truth is tied to the mathematician's *language*.

Every morning we would begin by reviewing the previous day's work. Almost always, Daniel would announce that our efforts had been worthless. Our notation was inelegant and clumsy. The lemma that we had laboriously demonstrated was trivial. Should our paper be published, serious mathematicians would regard it as a joke, and was there anything to eat? Sitting on the floor, with one hand shielding his eyes from the bright morning sunshine, Daniel would nibble moodily at the cookies that I produced, and then tamp tobacco into his pipe and smoke. His back hurt he said, and so did his knees, and what was He, a Jew, doing in exile from Zion, living by the waters of Babylon, and while he would gladly die for Israel, gladly die for anything at all, what did I think of streamlining our proof by the method of the elimination of quantifiers, and had I seen Mel Brooks on television the night before?

By and by he would perk up. Our paper became a monograph. When we had completed the details, we rewrote everything so that no one could tell how we came upon our ideas or why. This is the standard in mathematics. The question of the title came up. I suggested calling the paper *Logical Truth*.

"Too obvious," Daniel said.

"*Nu*, what then?"

"First thing, we have a picture of Rita on the cover."

Rita was a student of Daniel's with a sharp, appetitive face, and a great swelling bosom.

"A discreet shot," he went on, "Something from the neck to the navel. But tasteful."

"The title, *boichik*, the title?"

"What about *The Mystery of the Manifold Moistness*?" Daniel said reflectively. "Something wet in any case."

Truth, Theories, Models

Formal systems are the logician's stock in trade, however much like the rest of us they need to repair to their own limpid vernacular to get the point across. Some logicians, despite Gödel's theorem, stick doggedly to syntax and see in symbols only their shapes. This point of view yet dominates proof theory. Elsewhere, the semantics behind the syntax has long since returned to rosy life.

Mathematical model theory is not quite mathematical logic, but it is close enough to sustain an illusion of meticulousness. Model theorists study the relationship between various formal systems and the structures in which they are satisfied. These are generally set-theoretic, or, in any case, algebraic; when studied in isolation, they make up the domain of universal algebra. The logician looks backward to the formal systems, interpreting them now as *formal languages*. This gives to his subject its special and distinctive oomph.

And now for a number of delicate distinctions. From the perspective of *Penthouse* and its pets, mention of a model evokes an image of *une petite nymphe accroupie*, as a greater humbug had it. Among sociologists or social scientists – a big step down, of course – a model is a reflection of reality, so that by the time these people are finished with their labors, there are two of the thing and not one. What counts as a model in most places (sociology, urban affairs, psychology, political science) is reckoned a *theory* by logicians, and consists of nothing more than a set of sentences. The model is the world in which the sentences hold. The temptation is strong to dismiss such differences in terminology with a sniff of diminished interest. Still, a distinction discarded is a distinction denied, and an example of intellectual profligacy or simple bad taste. In calling theories models, scientists are left without a term to mark the important distinction between theories *and* their models.

Classical logic is bound up pretty much with the theory of the syllogism, and achieves its effects by treating names and nouns in abstraction from their meaning. It is thus that ALL MEN ARE MORTAL passes over to ALL A'S ARE B – an expansion of scope, a contraction of content. In modern mathematical logic, those dummy letters stay put, but the analysis is couched in terms of quantifiers and variables. Instead of saying, with Aristotle, that ALL A'S ARE B, the modern logician says that FOR ANY *x*, IF *x* IS an A, THEN *x* IS a B. With variables standing in for pronouns (or even names), he has the means to express cross-reference. The dummy letters themselves emerge in modern logic entirely as grammatical predicates.

By a formal language *L* I mean a *first-order predicate* language. Under the aspect of first-order logic, quantification extends only to individuals – one speaks of all *x*'s, but not the whole of anything else – sets or properties, say. The syntactic apparatus of *L* – its skeleton – consists of a list of individual variables *x*, *y*, *z*, . . . , the sentential connectives (*if-then*, *and*, *or*, *not*), the universal and existential quantifiers (*all* and *some*), the usual marks of punctuation (parentheses mostly, the logician's comma), and a collection of predicate variables of various finite ranks. A predicate variable ranked at 1 is a one-place predicate (*x boogies*); at 2, a two-place predicate (*x boogies with y*), and so upwards. Before he does anything else, the logician takes care to provide a completely explicit definition of the *well-formed formulas*. These are the grammatical strings of the language.

Under ordinary circumstances, words get their point and their poignancy by means of their interpretation, but the semantics of a natural language generally comes to nothing more than the informal declaration that there is something out there to which words refer and human beings respond. The logician does better by attempting less. Within mathematical model theory, the wonderfully rich relationship between a natural language and a natural world is replaced by the play between a formal language and a formal world. Let *D*, now, be a non-empty set of individuals. The choice of individuals is immaterial. A *relation R* of degree one on *D* is a subset of *D* merely (the set of all individuals in *D* who boogie, for example); relations of degree two are comprised of sets of ordered pairs (boogiers in *D* taken two at a time and in a specific order). A *relational structure M* is an ordered pair consisting, in the first instance, of a given non-empty set *D* (the *domain* of *M*), and, in the second, of various relations *R* defined on *D*. *M* makes for an *interpretation* of *L* if each predicate variable of *L* is mapped to a corresponding relation on *D*. Understood thus, *M* is a *model* of *L*.

The interpretation of a language within a model fixes its predicate structure. This is the first step in infusing the syntactic shell of a symbol system with meaning, and corresponds anatomically to the articulation of a skeleton. The semantic role played by the syntactic variables x, y, z, . . . is yet unspecified. Unremarkably, variables tend to be variable: Fx (*x boogies*) is true for some values of x, false for others. A technical trick is required to lend to the individual variables of a formal language a kind of temporary and artificial specificity. An *assignment* is a function that does just that: By mapping individual variables of *L* onto individuals in *D*, it fixes the interpretation of the variables in one stroboscopic flash.

The grand goal of formal semantics is to determine for every sentence the conditions under which the sentence is true – this for particular choices of *L* and *M*. Not every well-formed formula, however, makes a determinate statement: Shorn of its quantifiers, Fx hangs in midair. And yet the truth of sentences in which Fx finds itself embedded would appear contingent on the truth of Fx

itself. A closed circle in conceptual space now begins ominously to emerge. It is this consideration that prompts the logician to the definition of *satisfaction* as a relationship between assignments and well-formed formulas. When worked out in detail, the definition conveys nothing by way of surprise. An assignment satisfies a given formula if the formula holds in M when its predicate variables are interpreted as relations, and its individual variables pegged to individuals.

Truth is a concept that under the best of circumstances provokes a certain intellectual confusion, as when an undergraduate insists that some plain proposition is simply not true for *him*. However much I may pummel this poor character in class, I know what he meant to mean. In mathematical model theory, doubts of this sort disappear. Given the definition of satisfaction, the definition of truth follows at a single step: A sentence is true in M if and only if it is satisfied by every assignment, and false otherwise. If an unwholesome air envelops this definition, it is only because it runs against the grain of intuition to see truth cast in so ancillary a role.

●●●

The definition of satisfaction forges a connection between the well-formed formulas of a language L and assignments of value to its variables. The language in which the definition is itself framed makes up the working logician's linguistic vernacular. There is no reason why the relevant portions of this language cannot be recast formally along just the same lines of fastidious development that led to the elaboration of L. Assume this done. The result is a meta-language L^*. Considerations of elegance and economy might suggest that the introduction of one language to explain another is an exercise in unnecessary duplication. Why not let L explain itself? This is a good question, especially in view of the fact that whatever we know *of* English we know *in* English. In the case of simple first-order languages, no opportunity arises for the individual variables to refer back to the well-formed formulas. Gödel showed, as I have said (and said), that the natural numbers, when chosen as a domain of interpretation, could by a code be endowed with a self-referential voice of their own. The set of truths *in* a formal language corresponds thus to a set of numbers. But any arithmetical language capable of highlighting the Gödel numbers corresponding to its own truths is inconsistent. Such is the burden of Tarski's theorem, from which, incidentally, Gödel's incompleteness theorem ultimately follows. The upward ascent required for the definition of satisfaction is endlessly incomplete, like those irritating Escher-like pictures in which small fish are swallowed by fish just slightly larger in size. One longs for a fish large enough to swallow any other fish and too large to be swallowed in turn.

Elsewhere, the situation is different. The real field, for example, is not algebraically complete: $f(x) = x^2 + 1$ is a polynomial mapping with no real solutions at $f(x) = 0$. Like nature, mathematicians abhor a vacuum. This has prompted them to the creation of i, where $i^2 = -1$. The complex field $\mathbb{C}$ thus generated is in an irrefragable sense unique: Competitors are isomorphic to $\mathbb{C}$. The fundamental theorem of algebra shows that there is no need to go beyond $\mathbb{C}$, by adding a succession of exotic imaginary numbers to i, say. One step suffices. In looking at formal languages, it is, unfortunately, fish all the way.

Symbol to Symbol

The World, ancient scholars wrote, reveals itself by means of a secret system of signs. Whether the world reveals itself entirely or only in part, they did not say; nor, for that matter, did they offer much by way of a precise account of revelation. No matter. The doctrine itself suggests an associated image, with more than a little by way of contemporary appeal. To the right there is an independent external world; to the left, an articulated system by which the thing is represented. In that limit at which inquiry comes to a drowsy halt, the two come to coincide perfectly, with the system itself both true and complete. This makes for only one image but at least four metaphors: independence, truth, completeness and correspondence (or coincidence).

A natural language constitutes the largest and laxest system by which a world at large is duplicated at a distance. The image of the world set against the mirror of language (a large world, a small mirror, or the reverse, as in James Joyce) suggests a natural division of labor: The physicist gets the world or worlds, the philosopher, those languages by which it is depicted. This raises the question of *which* languages the philosopher needs to examine and how many worlds they describe. The American philosopher Nelson Goodman has argued that there are as many worlds as there are descriptions, which seems to me an embarrassment of riches. To describe the earth as a sphere for purposes of celestial navigation and as a point mass for purposes of celestial mechanics is to describe one thing twice. In his later writings, Wittgenstein suggested that not much appropriate to the great work at hand could be expected from the logician – rather an astonishing reversal for the author of the *Tractatus* – but evidently Wittgenstein's sense that philosophy stands or falls in the context of ordinary language had deep roots. Carnap mentions an early encounter with Wittgenstein in Vienna. Earnestly, the good natured Carnap expressed his enthusiasm for Esperanto – of all things! The very idea of an artificial language elicited from Wittgenstein, Carnap innocently remarks, only an expression of tremendous indignation, a kind of snort. The scene is incomparably droll: Carnap adverting to the ease with which he might communicate his research results to exotic logicians by speaking Esperanto; Wittgenstein muttering darkly about the organic roots of language. On the other hand, W. V. O. Quine, at least, has long seen in first-order formal systems a canonical notation almost Oriental in its purity; it is by means of *this* formal system, and no other, that the true and ultimate structure of reality may be discerned.

The proverbial man in the street or the physicist is inclined to argue, I think, that *they* get things right only when nature has impressed itself on language. That neutrinos have no mass is true *because* neutrinos have no mass. The conceptual connection goes from right to left. The image that corresponds most closely to this commonsensical declaration is that of the ordinary mirror, with a human language, or the human mind that informs it, acting as a smooth and undistorting pane of polished glass. Yet it is *also* true – that man in the street has been dismissed from this discussion – that what counts as a fact in general, or a neutrino in particular, is determined by a theory or a set of concepts. The conceptual connection goes from left to right. The image that corresponds most closely to this counter-intuitive declaration is that of motion picture projector, throwing images onto a silver-studded screen. In mathematical model theory, common sense and counter-intuition come to coincide. A first-order language reflects perfectly a first-order world; a first-order world arises as the result of the projection of a first-order language onto an abstract space.

The great clarity of mathematical model theory has led a number of philosophers to the conclusion that, in general, worlds arise only in response to words. Nelson Goodman has, for example, committed himself cheerfully to a cagey variant of *irrealism*; Richard Rorty has argued that even the very contrast between words and the worlds to which they refer represents an illusion. There is no perspective, so his argument runs, from which the world at large may be described as it *really* is, and thus – the argument now takes a second dainty but dramatic step – no world at large either. The world is *entirely* a matter of visions, versions, and views – commodities, of course, which among philosophers are never in short supply.

I mention these dialectical twists and turns not to endorse any of them, of course, but to establish a moral of sorts. Analytic philosophy has long been vulnerable to the attractions of a certain form of renunciation. The realist holds for the existence of an external world and looks to physics to see its reflection in the mirror of theory. Hearing him out, the logician observes that the world is as much made as mirrored. And the irrealist argues that so long as worlds are made, there is little sense to supposing that they are mirrored at all. Beyond any of these positions, acting as what mathematicians call a limit point, is *scepticism*, the doctrine that neither worlds nor their mirrors exist. Each position leads imperceptibly to the next, striking evidence for the sheer slipperiness of sin.

The Range of Light

In the spring of that year, after many weeks of work, Daniel and I decided to hike the high Sierra Mountains – the range of light. My own experience with mountain climbing had been pretty much limited to the Catskills in New York State, which are lovely, history-haunted hills, and the White Mountains of New Hampshire, rather a sullen, sinister, and dangerous range, actually, despite their low elevation.

We knew nothing of serious mountain climbing. For a few days we thought to attend to our physical condition. We embarked on a running program that was to reach five miles in one mile steps – a mile the first day, two miles the next, and so on. Daniel was much concerned to obtain the fabled runner's high, but after we had trudged that first half mile and more or less collapsed, he suggested a regimen of angel dust and cocaine instead. "At least junkies don't need to buy expensive shoes," he said, looking sorrowfully at his new Adidas. "Just a spoon and nine inches of rubber hose, and you've got the same equipment as the very best."

By and by, we found ourselves at Tuolome Meadows, the John Muir Trail ahead of us. We both carried heavy, inexpertly packed knapsacks. For some obscure reason, we thought that we might fish along the way, and had affixed a pair of bamboo poles to the outside of our packs. The packs were so heavy that neither of us could swing them onto our own shoulders unaided. When we had both finally gotten into our harnesses, we stood there swaying in the harsh, glorious mountain light, everything aching all at once. We began to trudge, with Daniel moaning to himself at every step. On the map it looked easy. We were to hike ten miles to the base Mt. Lyell and then ascend the peak by its north face and descend to the south, but after an hour or so, we realized that we were walking in the wrong direction.

The heat was remarkable. We had gone no more than five miles when we decided to rest and do some fishing in the bright, babbling brook that ran through the meadow. Somehow or other, we managed to assemble our fishing poles and fasten hooks to the gleaming, silvery, very fine line. On our very first cast, our lines simultaneously became snagged in an overhanging tree branch. No matter what we did, or how we struggled, the lines became hopelessly more snarled until finally, in desperation, we sliced the poles from their lines, and solemnly assembled all of our fishing gear and threw it into the creek with an angry mutter.

We camped that night by a ravine on the side of a mountain, built a huge fire – visible, I am sure, from distant planets – dumped virtually all of our dried food into a pot, set out the bottle of Johnny Walker Red that we had thought to bring along, and ate and drank ourselves into a stupor. We fell into a drunken discussion of how best to calculate the distance from our camp to the bottom of the ravine.

"Simple," Daniel said. "Compute the integral." The idea was that we would drop an object into the ravine, and measure the time it took to reach bottom. Its acceleration we knew by elementary physics. The distance that it had travelled we could then calculate by means of the calculus. Gesticulating with our guidebook, which contained all of our maps, Daniel managed in his enthusiasm to hurl the whole thing down into the chasm, an expression of bewilderment and dawning dismay on his expressive face.

We awoke the next day with the kind of particularly vicious hangover that high altitudes provoke. Above lay the snow, below, the valley through which we had hiked. We thought briefly of turning back, but we had reached the halfway point, and the snowy mountain peak seemed more inviting than the dry, hot, buggy valley. When it came time to put on his expensive North Face boots, Daniel discovered that his feet had swollen so badly that he was unable even to mash his toes into the box of the boot. All that morning, he climbed barefoot in the snow, trudging upward like an anchorite.

Somehow or other, we made it to the top of Mt. Lyell, where we urinated ceremoniously on the very peak, the spring wind raw, alien. Daniel, or course, found himself facing the wind and managed to drench us both.

Two days later we stumbled from the woods and emerged on a state highway in Nevada, before a diner with a tattered sign marked EATS.

The rosy-armed woman behind the counter looked us over for a moment.

"You boys look like animals," she finally said in a calm, kind voice.

The Ghost within the Machine

Alan Turing came to Princeton in 1936 to study with the logician Alonzo Church (the same Church with whom I had studied, then slim as a steeple). He was nineteen at the time, as elegant as a slept-in shirt, and a classic British ectomorph – I am judging from his photograph, which depicts him in running shoes and shorts, his head at an awkward angle, *les yeux perdus*.

In his temper, Turing belonged in the company of history's great-hearted cranks. This is an image that expresses his inventiveness, his curiosity, even his perseverance. It fails to do justice to the striking mathematical depth of his thought. Along with a handful of other mathematicians and logicians, Turing worked at the very margins of the mathematical experience, unsure whether he would ultimately fall from that perilous edge and vanish into the void.

To the question of how thought proceeds, the normal response, I think, is by a mixture of insight and intuition – a kind of forward lurch that remains phenomenologically insusceptible to specification. Consciousness records a fast blur of tension and release. It is only when thoughts are expressed in language that an activity of the intellect appears amenable to decomposition – thoughts are then mapped to sentences; sentences to words. What remains is thus a finite alphabet of *symbols*.

A machine, I have argued, is the most general of devices taking inputs to outputs by means of a set of states. The human subject begins with words and ends with words. In going anywhere, he has only words to go on. But in passing from one set of words to another (from hearing what he says to saying what he hears, for example) a human agent must make use of a purely physical object. In carrying out any particular computation, Turing reasoned, the brain occupies a particular neurophysiological configuration or state; these the neurophysiologist identifies with a distribution of neurons or even a collection of chemical pathways. And, yet, the human brain is finite. So too, then, are the states that it may occupy. The identification and analysis of the brain's neurological states Turing quite properly thought a task for the biologist. The mathematician need only attend to their *existence*. By means of a great dreamy imaginative leap Turing merged these observations with that definition and concluded that thinking is purely a mechanical process. Having reached this high ground, he wondered whether the machine that instantiated an act of thought need be the human body. This very natural and dramatic question suggested to Turing the idea of a *Turing Machine*.

A Turing machine is designed to manipulate the elements of a set of symbols – words, say, or numbers, or letters. These constitute the machine's alphabet; they are displayed for the machine's examination on a tape divided into squares. The tape stretches infinitely in both directions. Above the tape is a reading head that serves to scan the squares. It may itself move one step to the right, or one step to the left. It may not move at all – an ancient, sphynx-like eye. Symbols from the alphabet the reading head inscribes on the tape, one symbol to a square; those symbols it no longer requires, it extinguishes by erasure.

Like any other machine, a Turing Machine is designed to *do* something. Confronting a set of symbols, its life is pretty much a matter of their transformation. It is thus a device of lunatic literalness, and indistinguishable, on this score, from the rest of humanity. At any given integral instant, a Turing Machine is capable of occupying one of a finite number of internal states: Its *behavior* is completely determined when two temporal streams are fixed. In the first, the reading head changes its position on the tape; in the second, one state gives way to another.

Having manipulated variously the symbols that it examines, a Turing Machine has transformed an *input* tape into an *output* tape.

There are certain concepts within mathematics that carry their significance on their wrist like a great tattoo. With other concepts, the significance takes some seeing. In what follows, I outline the construction of a specific Turing Machine capable of adding any two natural numbers. *Any* two, note. The machine's alphabet consists of the symbols 0 and 1. A natural number is represented by the machine as a string of $n + 1$ consecutive 1s – 4 is thus simply 1 1 1 1 1 . At any time, this machine may occupy one of six internal states, which I designate by means of the symbols s_0, s_1, s_2, s_3, s_4, and s_5. The symbol # indicates that a square on the tape is blank; L and R, that the machine is to move either to the left by one square or to the right. At HALT the machine stops. There are eight lines of instruction:

1	s_0	#	s_0	#	L
2	s_0	1	s_1	1	L
3	s_1	1	s_1	1	L
4	s_1	1	s_2	1	L
5	s_2	1	s_2	1	L
6	s_2	#	s_3	#	R
7	s_3	1	s_4	#	R
8	s_4	1	s_5	#	HALT

The first line has the following meaning: If the machine is in state s_0, and is examining a blank square, it is to print the symbol # on the square, and then move to state s_0. (It remains in the same state.) It then moves one square to the left. The remaining lines are read and understood in the same way.

Even the reader to whom mathematics is an affliction should appreciate the stunning power that mysteriously inheres in symbols. Here on the printed page, in a physical arrangement of ink (and that a concession only to the limitations of the *human* memory), is machinery enough to carry out the addition of any two numbers – stunning evidence, if any were needed, that the laws of thought owe little to the laws of physics, and represent, instead, a world in which concepts, like dreams, move entirely according to a logic of their own.

In the voluptuous generality of its conception, a Turing Machine is capable of standing in for any number of other concepts. From one point of view, it acts as a very model for the activity of computation; from another, it gives elegant content to Hilbert's notion of a formal system. But Turing's description of his Turing Machines also laid the foundations for the development of computer science, and so formed one of those landmarks in thought by which the past and the future are sharply divided.

Hilbert thought of mechanical reasoning in terms of a formal system whose elements were symbols and their rules. Turing thought in terms of an abstract machine. The formulations are quite different. The concepts are the same. Every formal system may be represented as a Turing Machine. Every Turing Machine embodies a formal system. In one of those inexplicable accidents of academic life, the American logicians Alonzo Church and Emil Post, following quite different lines of reasoning, defined a variety of abstract objects that in the end found expression as Turing Machines – an indication in such matters that the very same idea was being given inessentially different formulations. This prompted Church to argue that only one idea had been defined – *effective computability*.

It was Hilbert's idea to make mathematics mechanical by the double action of first withdrawing meaning from a set of symbols, and then reducing inference to a series of discrete, combinatorial steps. Gödel put an end to the

Hilbert Program. Arithmetic is incomplete; its consistency cannot be demonstrated within arithmetic itself. This served to show that some truths must inevitably wriggle through any axiomatic net. Wriggling thus, they escape completely from mechanistic constraints. Gödel's theorem (and Tarski's theorem as well) established that the truth of an arithmetic proposition and its provability within a formal system are separate matters. There yet remains a single, large-hearted question about the scope of purely mechanical concepts within mathematics. In the propositional calculus, the tautologies and the theorems coincide perfectly. This circumstance makes it possible to decide, with respect to an *arbitrary* proposition, whether it is a theorem. The mathematician need only establish whether it is a tautology. Having established *that*, a procedure exists for finding its proof. The problem in outline suggests the ping-ponging of a pair of questions: Is an arbitrary statement of a formal system a theorem? If it is, what (or where) is its proof? The questions taken together constitute the *decision problem* for a formal system. In the third part to his (horribly unlucky) three-part program, Hilbert asked whether in general a decision procedure existed by which the decision problem could be settled. By a decision procedure, he meant an algorithm. That concept, of course, collapses into the concept of a formal system or a Turing Machine. In any event, to conclude on a negative note, the result of Turing's investigation was negative. Gödel showed that the truths of arithmetic did not all follow from the axioms of arithmetic. Turing demonstrated that the decision problem was insoluble as well. An element of thrilling and unassailable waywardness yet attaches to mathematical thought.

An Attitude Adjustment

I met Marvin Minsky only once. I had been invited to deliver a paper at a NATO conference in the south of England. I arrived in London in the late afternoon, punchy from jet-lag, hungoverish. I had not shaved, nor washed. I took a train to the south and spent the hours that it took to get there gazing at the humid-looking sky and those lush, heavy English fields in which cows graze underneath apple trees. The conference itself was held in an outrageous country estate calculated to suggest the imminent appearance of Sebastian Flyte: I expected to see him peep from the hedges, sniffing a rose, with shaggy Charles Ryder a step behind him. Dinner that night was presented in a long gloomy hall. There were, perhaps, twenty of us in all – academics from the United States and England, with a few ill-at-ease professors from Finland or Greece there to give the proceedings an international flavor. I had given up cigarettes some days before. Everyone at my table smoked, with one woman that I recall, a psychologist, I think, dragging at her French cigarette with such voluptuous abandon – her thin cheeks hollowing to suck the smoke in – that I could have throttled her on the spot. Only the prospect of spending several years in a provincial English prison, eating English food, dissuaded me. Afterward, the men retired to smoke cigars and play snooker, a game much like pool except that it is played on a table half again as large as a pool table. I tried my hand for a few minutes and then walked outside. On the misty hills, songbirds swooped and cheeped, darting this way and that to catch insects. I wandered about for an hour or so, sinking slowly into depression – *platzangst* – and then retired for the night to a room that was guarded from the outside by a suit of armor.

The talks began the next morning after breakfast. The first speaker walked diffidently to the platform. He was very glad to be there (or here), he said. A great American virtuoso of artificial intelligence, he had programmed a computer to diagnose diseases of the blood. Now he was engaged in explaining how the program worked. At a point halfway through his talk – judging the matter by the number of unread papers on the podium – he caught a frog in his throat. It must have been the size of a grapefruit. No sooner had he cleared his throat with a great wet snort than the thing would tickle him anew until he was forced by circumstances to wave his arms impotently and promise to continue his address at some other time. The moderator of the session was Michael Arbib, a computer scientist whose name I had known at Stanford. He sat deeply slouched at the head of the table, staring at the ceiling and twirling a paper clip in his fingers. The man sitting next to me, I remember, was a mathematician from Michigan; he seemed to have dressed in a shirt several sizes too small for his neck. As the first speaker gasped and sputtered on stage, this character, sitting quietly, changed color sympathetically, like a chameleon, his head and face turning an alarmingly deep shade of magenta, the pair of them, the speaker and my table-side companion, apparently moving inexorably toward cardiac arrest.

It was my turn to talk. Now a secret must be imparted. I had nothing whatsoever to say. One never does at these meetings. I had prepared myself for the occasion by purchasing a very natty new suit (gray, with a thin blue piping). Academic life has always revealed itself to me in terms of the opportunities it affords to display my wardrobe. I was not five minutes into my speech when a short, squat, bald figure – Marvin Minsky – popped up and in his stevedore's voice began obstreperously to list at least a dozen perfectly sensible objections to my remarks. I quieted him down only with difficulty. Two minutes later, he was up again, waving a ballpoint pen in the air, gesticulating madly. From the far corner, Michael Arbib snickered to himself in a gloomy, abstracted way. I paced back and forth, dying of hatred and irritation. I was about to suggest something unprintable – the words were just forming on my lips – when strangely enough Minsky lapsed into morose silence, twitched once or twice, and then shambled off intellectually on some other private mission.

I finished what I had to say in short order and mooched a forbidden cigarette from the psychologist whom I had previously wished to throttle.

Later, still stung, I heard one of the Bach fugues from the *Well Tempered Clavier* being lovingly mangled on a piano hideously out of tune. I opened the door silently. There was Minsky, dressed in a red turtleneck sweater, the sleeves pushed above his hairy forearms, his convex forehead beaded with perspiration, swaying as he played, and singing.

The Machine within the Ghost

The modern digital computer – the so-called von Neumann machine – was born of an apparently innocent flirtation among a number of abstract ideas – intellectual polygamy, to sort the thing out by sexual roles. In the seventeenth century – another range of light – Leibnitz wrote of a universal calculating system by which any conceivable intellectual dispute could be mechanically resolved. This subversive and astonishing idea remained in western consciousness during the centuries that followed, a kind of shaggy nightmare. On the other hand, there are those very literal machines that lie littered in the prehistory of the digital computer: The Jacquard Loom, the cotton gin, and even some great brass and chrome monstrosity constructed by the cranky Englishman, Charles Babbage – the so-called analytical engine.

Whatever its ancestors, the computer is itself a machine made of matter, an object thus in the physical world. The first real and recognizable computer was the ENIAC, which was constructed after the Second World War at the University of Pennsylvania. The ENIAC performed calculations at the rate of twenty thousand multiplications a minute – slow by present standards. The ENIAC, in turn, begat the EDVAC and the EDSAC – genetic improvements on the ENIAC and a clear example of natural selection at work. Both machines accepted stored programs. The development of the transistor at Bell Laboratories in 1947 made possible the next generation of machines. The integrated chip followed, and then the large-scale integrated chip; and then the very-large-scale integrated chip. By conventional reckoning, four generations have now passed. It is thus that the computer is descended on its several sides. The Japanese aver that they are involved in the construction of the fifth generation of computers. These are machines designed to speak in Japanese. They are said to evince a desire to eat raw fish.

For all its evident complexity, the modern digital computer is essentially designed to manipulate strings of binary symbols. To this end, every von Neumann machine consists of four parts. There is the central processor, which stores and controls the machine's program; the memory; the arithmetical unit; and a device to handle the ingestion and excretion of information. At its very bright, very simple heart, the computer contains a switch in silicon that may be set in one of two positions. What makes for congruence between binary numbers and binary switches is simply that any binary number may be represented by switches set to either ON or OFF. The number 100010 corresponds to the following setting: ON, OFF, OFF, OFF, ON, OFF. And any number may be expressed as a binary number.

There is a close family connection between a digital

computer and a Turing Machine. The Turing Machine is an abstract computer whose vast strength lies in its infinite memory. Existing as it does as an abstract object, the Turing Machine collapses the distinction between hardware and software. The Turing Machine thus serves as a model for both the computer and its program. Whatever it is that any particular computer may do, a Turing machine may do as well, although not as fast. This lends to the Turing machine a kind of brooding inscrutability. Looking at the instructions for a Turing Machine, the logician cannot in general determine whether the machine will ever reach an output, or whether the device will encounter a kind of mental block or loop endlessly, drunk with its power and inconclusive. These are logical limitations. A real computer, or its program, is bounded in space and time and must make do with a memory that comes to an end. The most carefully composed program is yet apt to torment and baffle its creators – by sending the bank's statements to the Aleutian Islands or peevishly triggering a nuclear alert. These are empirical limitations and suggest somehow the cheering possibility that the digital computer – its formal structure, in any case – is as inadequate as the rest of us.

Blind Ambition

Like Rasputin, certain ideas simply refuse to die, and, indeed, seem often to exhibit an awkward sense of their own lunatic vitality at precisely the moment that their assassins have most recently emptied their revolvers. There is, for example, artificial intelligence.

During the early years in the development of the theory, with the red sun of ripe hope rising, theorists concentrated chiefly on the construction of computer programs that might simulate very specific intellectual skills.

The programs were of two severely separate sorts. Some were simply formal systems designed to proceed in sequence to a specific end. These typically involved the computer in an extensive search of some domain; in their intellectual structure they did not generally go beyond an order offered to Rover to fetch. Others incorporated rules or rubrics that seemed tolerably close to the strategies invoked by human agents. These programs embodied a strong and definite hypothesis about the world external to the computer. Yet, by common consent, the theories that resulted achieved nothing more than the most limited life, and if they resembled anything resembled mostly those patients who persist in smiling from their iron lungs.

The rising sun has turned from red to raw. There are always philosophers who regard the embarrassment of others as an intellectual invitation. In *What Computers Can't Do*, Hubert L. Dreyfus examined in turn the claims for artificial intelligence in language translation, chess, and a variety of cognitive tasks. He reported that the field gave every indication of "diminishing returns, disenchantment, and, in some cases, pessimism." Needless to say, the expression of this position did not endear Dreyfus to his colleagues working on artificial intelligence, who when pressed on this point would be heard to offer the opinion that it is better to light a candle than curse the darkness.

Me? I've always been in favor of cursing the darkness roundly, so this aspect of Dreyfus' thought appeals strongly. To a certain extent, I am obliged to report, Dreyfus, writing in the early 1970s, wrote too soon. The astonishing development of the computer has made possible a more sophisticated attack on the problems of artificial intelligence. I am thinking now not only of matters of size (smaller) and speed (faster). There are also new programming languages and entirely new techniques of parallel processing. In many areas of physics, computation has replaced analysis. Equations that defiantly resisted analytic techniques for decades now submit docilely to simulation. To see a computer simulate the solution to some hideously complex set of equations, and then flamboyantly exhibit its calculations graphically, is, without doubt, to be a witness to a display of some sort of intelligence; for anyone trained in old-fashioned analytic techniques, the result is very much like seeing for the first time parts of a landscape under strong light. This is always an unsettling experience (especially in the graphic arts). It is entirely possible, I suppose, that, as a discipline, artificial intelligence will proceed in small stages until at some point in the next century or some century beyond, computer scientists and philosophers will be in a position to point shyly at their discipline and announce that there it is – the whole of human intelligence on a chip.

I doubt that this is how things will turn out. Science generally proceeds by means of strong theories. When it comes either to human or artificial intelligence, the theories on hand are weak, partial, inconclusive. Beyond this, there is a more general point. The theory of artificial intelligence is a discipline committed to a single perspective. Intelligence, so the theory must run, is an activity achieved by passing through a number of perfectly determinate steps. This is not merely a practical perspective – one evoked by the demands of computer science. It is theoretical as well. Church's thesis that effective computability *exhausts* the informal concept of an algorithm means that whatever may be executed by any new machine or language may be executed by a Turing Machine as well. It is possible thus to repair to certain very familiar intellectual acts and ask whether there is any prospect that they will *ever* admit of sequential simulation. It is precisely these familiar acts or attitudes to which Dreyfus attends. The issues remain wide open and full of sin.

Take pattern recognition, for example. In moving

through the world, a human being quite unconsciously comes to sort and classify the objects of his experience. It is this ability that enables a man to tell one thing from another – a circle from a square, the letter A from the letter O, the face in the mirror from the face in the crowd. Unless a machine be given such powers, it is unlikely that it will ever exhibit anything much like intelligence, however well it may execute a variety of routine, pre-set tasks. The Grand Master, looking at the board, sees the chess pieces and the squares in terms of strategies: These are patterns over time. Only the hopeless amateur examines each piece in isolation. In translation, the skilled bilingual accepts the data in clumps, forging an eerie unity out of what to the novice appears as a string of words. Now a computer programmed to recognize a pattern must proceed according to a set of rules. These are designed to associate a pattern with some fixed set of features. The features are fixed because the computer is finite. They are features because there is nothing else.

This suggested to the bouncy and irreverent Dreyfus two problems of uneven difficulty. In teaching a computer to read the characters of a written language, it is plainly possible to compile a finite list of features that distinguish a printed A from a printed O. In this limited sense, a computer may be given quite rudimentary powers of pattern recognition. But type fonts vary. There is italic, cursive, roman and gothic. Characters may be written by hand. To accomodate these variants, that list of significant features must grow explosively. Quite soon, computer memories become exhausted. Yet human beings do manage the whole thing. The ordinary literate adult finds it easy to read characters in any script. Those with which he is not familiar – German Gothic, say – he quickly learns. How?

I mention this problem to suggest the nut within the shell. A machine that is able to distinguish a circle from a square has acquired a cognitive skill. This skill it has achieved by proxy. Those features it is now capable of recognizing have been incorporated into its program by human intervention. But human beings generate pattern-recognition features on their own, quite without instruction. Even the week-old infant, helpless in almost all respects, is capable of detaching Mama from the background in which she is embedded. Unless otherwise instructed, the computer tends to merge Mama with the rest of its perceptual field.

Human beings, Dreyfus observed, allowing his own argument to expand like a wave front, appear to possess groups of psychological attributes that do not suggest the operation of a finite set of rules at all. These attributes he lists in pairs of opposites – fringe consciousness as opposed to heuristically guided searches, for example. A heuristic search is what the computer undertakes in chess when it examines a tree of moves and, by various pruning strategies, attempts to select the next best move. Fringe consciousness is the melodious awareness of pattern that suffuses the master or the madman as they contemplate *their* next move.

Then there is the distinction between ambiguity tolerance and context-free precision. Here the relevant examples arise when a computer is instructed into the mysteries of a spoken language. "Stay close to me," when taken simply as a sentence, may mean any number of things, with closeness measured by millimeters or miles. The interpretations obviously vary with the context. The contexts vary indefinitely and are limited only by the imagination. How can a computer be given a list of relevant contexts when no one, certainly not the programmer, has any idea of the extent or nature of the list?

I read all this with a great deal of sympathy. Other than observing that human intelligence appears to be a whole that is not diminished by any subtraction of its parts, I am unable to say what it all means. In explaining his own program in philosophy, with perhaps a touch of mournful embarrassment, Dreyfus turns for assistance to Heidegger:

> Equipment has its place or else it "lies around": this must be distinguished in principle from just occurring at random in some special position. . . . The kind of place which is constituted by directions and remoteness (and closeness is only a mode of the latter) is already oriented toward a region and oriented within it. . . . Thus anything constantly ready-to-hand of which circumspective Being-in-the-World takes account beforehand has its place. The "where" of its readiness-to-hand is put to account as a matter for concern. . . .

Friends of mine, it is true, in reading this passage, and many others like it, have often responded after a few moments of introspection with a lively "Hey," as if a pure white light had entered their consciousness. I remain convinced that they are posing merely, and that the whole thing – the positive Dreyfus, the passage that I have just quoted, and the pose I have described – is nothing more than an elaborate joke.

●●●

I believe. I want. I do. What could be simpler? Intelligence is the overflow of the mind in action. In dreaming or desiring, on the other hand, I occupy a world bounded entirely by memory, meaning, and belief: I need *do* nothing. That overflow is entirely internal, as in an erotic dream. In either case, our intelligence is *directed* toward specific objects or states of affairs. I believe – what? That *Stearsil Starves Pimples* or that *Pepsi is the Choice of a New Generation*; I desire – what? That the young Sophia Loren might step smouldering from the television set for perhaps an hour or that I might win a MacArthur Fellowship (the academic equivalent of the Irish Sweepstakes). What *I* believe (or desire) and what *is* believed (or desired) are connected by something very much like an intentional arrow, a kind of miraculous metaphysical instrument. The relationship between my thoughts and their objects is thus strange from the first. But this relationship between what I think and what I think *about* is duplicated in language itself: Like the thoughts that they express, the sentences of a natural language transcend themselves in meaning.

In seeing things from a first-person stance, with the entire world revolving around my own ego – a kind of Ptolemaic system in psychology – I direct the arrow of intentionality from the inside out, infusing the objects and properties of the external world with all of the significance that they ever possess. I assume, of course, that others do as much. Read forward, the arrow of intentionality goes from what I feel to what I do; read backward, from what is done to what is felt. The sense that we are all in this together arises only as the result of a supremely imaginative kind of back-pedaling; the interpenetration of two human souls, when it occurs, is wordless.

There is more. Each of us acts in the world as both subject and object: We do, and things are done to us. In moving away from the lunatic solipsism in which my ego exists in the absence of all others, I endow those human beings in my own perceptual ken with more or less the same cognitive states that I myself enjoy. This is the basis for a sense of sympathy. The endowment itself, I presume, may be reversed, as when I myself figure in someone else's awareness as an imaginatively constructed subject of experience. But here is a queer, artful point. The inferences that I make about others, others make about me. My inferences about others I cannot verify, but their inferences about *me* represent something like the backward wash of a familiar wave. A subject acting simultaneously as a psychological object enjoys a unique Archimedean perspective on the system of inferences by which mental life in the large is constructed.

This confluence of circumstance suggested to the American philosopher John Searle a very deft argumentative maneuver, something akin, really, to a movement in *judako*. His arguments were prompted by work undertaken at Yale by the psychologist Roger Schank. Like many other American theorists, Schank has approached the problem of artificial intelligence with a kind of bluff, no-nonsense sense that getting a machine to understand something is a matter of attending to the details in a patient, straightforward way. In a photograph at the back of his book, *The Cognitive Computer*, he stands with his arms folded over his ample belly, scowling directly into the camera, an expression of earnest ferocity on his face, as if to suggest that by the time *he* got through with them, those computers of his would either shape up or ship out. His aim, as he explains things, is to teach the digital computer to comprehend simple stories of the sort that might be told to children. The exercise is set out without irony. The education of the digital computer in this regard commences with what Schank calls a *script* – a kind of running, rambling background account in which the saliencies of various stories are set out and explained. With the scripts in hand, the computers are prepared to make sense of what they read. They are then interrogated with a fine eye directed toward telling whether they have understood what they have absorbed. In fact, Schank's machines *do* get quite a bit right; the record of their conversation is admirable, and the unbiased reader often has the feeling that just possibly he is reading something strange and remarkable.

It is against this conclusion that Searle has set his face. It is a simple fact, Searle begins, that he is utterly ignorant of the Chinese language. Suppose that he were to be locked in a room with a large sample of Chinese script – the samples, say, arranged on cardboard sheets. Now imagine that Searle were to be given "a second batch of Chinese script together with a set of rules for correlating the first batch with the second." The rules are in English. A third collection of scripts is presented Searle. And another set of rules. This makes for three separate sets of Chinese symbols and two sets of English rules.

From Searle's point of view, the material he confronts is an incomprehensible jumble. From the outside, where sense is made of all this, those Chinese symbols have a specific meaning. The first corresponds to a general script – the sort of thing that a computer would need in Schank's setup to make sense of a story. The second is actually a story in Chinese. The third represents a list of Chinese questions. From time to time, those questions are presented to Searle with a nudge and a wink and a tacit request that he say something. In answering, Searle consults his set of rules. The two sets enable Searle to match the questions to the story by means of the background script. In this respect, Searle remarks, he is precisely in the position of the digital computer.

But (a very excited, explosive *but*!) under such circumstances would there be any inclination to say that a subject so situated understands the meaning of the symbols he is manipulating? An *observer* might come to this conclusion. Put a question in Chinese to this character, after all, and he answers in Chinese. Yet this is not at all how Searle himself sees things. Whatever he may be able to *say* in Chinese, he remains confident that he *understands* nothing of what he has said and is prepared to champion his ignorance defiantly. Some great notable aspect of what it means to understand a language has simply been overlooked.

For the most part, computer scientists have tended to ignore Searle's argument and the point of view that it represents. It had long been known in science that you cannot beat something (a research grant) with nothing (a destructive argument), and what Searle had to offer them was nothing at all. Analytic philosophers responded promptly to Searle. The results are confusing. A great many superbly confident rebuttals appear to contradict one another. As for myself? When pressed on the point, I tend to run my hands through my hair or tug mournfully at my ears, gestures I am convinced that suggest that I have something tack-sharp to say were I willing only to say it.

Late Lunches

I met Hubert Dreyfus through the suave intervention of my friend, the mathematician Gian-Carlo Rota, and John Searle through the intervention of Hubert Dreyfus.

We saw each other socially several times at Berkeley, Dreyfus and I, generally at one of a number of hideous restaurants. It is a characteristic of restaurants in Berkeley that of any two each may be considered inferior to the other. Dreyfus was short and brassy and explosive. He had flaming red hair. I liked him for his intellectual openness and his willingness to address large issues without worrying overmuch whether he understood them.

One day we met in the faculty cafeteria. Dreyfus and I sat awhile, talking of nothing much in particular. Then John Searle came bouncing into the place, seeming to use up the available oxygen in a single gulp. He was a short, stubby, combative figure, dressed in khaki and brown work boots, a belt slung underneath his stomach. He had brown hair and brown eyes in a wide, somewhat irregular face that might have been perfectly at home in a barracks or a barroom, and the easy shambling power of someone born with authority but not grace.

He wanted to know what I thought about something. I cannot remember what.

"Well, what'ya got?" he asked.

"What do you want?"

"A knock-out argument."

Searle evidently thought of himself as something of

a philosophical prize-fighter, much in the business of delivering knock-out blows to a variety of preposterous theories. His article on artificial intelligence, he was convinced, had finished up the field once and for all. I thought to remark that had I the sort of argument he wished, whatever the topic, I would certainly have used it myself.

It was very curious sitting there in Berkeley and talking about artificial intelligence or linguistics or theories of free will and action. Thirty miles to the east, in Silicon Valley, a vast corporate culture was at work on precisely the sort of problems that Dreyfus and Searle were convinced could be neither settled nor solved. Our conversation reflected a touching faith in the power of the word.

But within minutes, to be fair, Searle lost all interest in what he had been saying and began roaming the room impatiently with his eyes.

Sermon for the Day

Human happiness is a matter of a short list: The right Rolex (steel and gold, *bien sûr*); a Mercedes Benz 280 SL, the kind that is no longer made and that features a short, snappy snout and a flowing back with a womanish *ensellure;* a gold and chrome VCR, suitable for late night pornography, its face glowing luminously; and, retreating now to the real world, a pliant blonde with melting lips and antelope knees.

I know, I know. Eastern mystics have long urged that the path to salvation, which, I am afraid, passes uncomfortably over a razor's edge, involves the stern renunciation of material possessions. But there it is – I am interested in the Absolute only to the extent that in getting there it is possible to drive. During the dark night of my soul – between two and four in the afternoon, as it happens – the voice that I attend to is forever whispering *More* to me in a throaty growl, rather than *Less*.

In looking over my own list, or one offered in an edition of *Playboy* (an article on the girls of Peru running amicably beside an interview with a rock star – *I dunno, really, the music isn't the main thing, it's the money*), I am struck by the extent to which so much of what we want (that blonde excepted) is bound up with things we make. Faced with the biological imponderables of life itself, there is some purely human craving, as Yeats observed, to take refuge in the inanimate – things that may break, but do not decay or droop or succumb to wretched excess or acquire any of a number of loathsome diseases.

The class of artifacts covers anything that is man-made, but machines are artifacts in a doubled sense. They are man-made, of course, and in this sense they do not grow or come gradually into being or fade out of existence; but machines are artificial – *inhuman* – in a deeper and more disturbing sense. What a machine does is fixed by what is done to it and its set of states. Once embodied in hardware, a machine is an object like any other, and suffers change. But its *essential* aspect is indifferent to time. A machine exists in a monstrously pregnant present.

Mechanism in natural philosophy is, I have suggested, the doctrine that in the homely pump (these examples may be varied at will) one has an explanation of the human heart. *Reduction*. It is very easy to pass from the declaration that the heart is, among other things, a pump, to the far stronger declaration that the heart is *nothing but* a pump, and hence itself a mechanical object; the brain, Marvin Minsky once remarked, is nothing but "a meat machine". *Generalization*. Having established that the heart *is* a pump, physicians are disposed to get rid of the original in favor of some frank substitute. *Elimination*. In reduction, generalization and elimination, one has the characteristic mental movements of much of twentieth-century thought.

And yet, I am inclined to say, the beating human heart is greater and grander than a pump. The line between what is natural and what is mechanical remains resistant to trivial erasures; even the physical heart, the thing in flesh, has its store of secrets and cannot completely be represented by means of its dynamical states. Neither can the human soul. Why this should be, I do not know.

End of Sermon.

PART IV

Alpine Weather

I came to work for Flyte and Company during a hot, sunny, New York June, and left covered by a cold cloud just nine months later. Alpine weather, I suppose. On my first day, I was given a lovely, mint-green American Express card, a packet of company checks, an appointment book of polished, Vermeer-Red leather, an elegant fountain pen, which commenced promptly to leak and never thereafter ceased, a set of notebooks containing brief biographies of the firm's senior partners (all of whom lived in Connecticut and had wives named Flo – a flock of Flos), an account, bound again in leather, of the firm's history and most notable successes (*Improving Profit Potential at the New York Herald Tribune*, a title that I remember), and, finally, a collection of embossed business cards, with my own name in raised boldface, the legend *Consultant to Management* in discreet italics below.

Some weeks before my arrival, the grizzled old chief of a major firm had been unceremoniously ejected from his office by a crowd of corporate revolutionaries. The new chief, some dark figure whose name, I think, was Armbruster, had evidently spent some time on the golf course with Roger X, a Flyte director, and an elegant, ineffective, dreamy character, with soft white hands, silver hair, and a deep ridge of unhappiness between his furry, brushed eyebrows. After arriving at the office in a maroon limousine, it was X's habit to repair to the executive washroom for morning ablutions; he then spent the rest of his day sitting at his immense walnut desk, straightening the starched cuffs of his Sulka shirt and staring into space. He was not a failure, plainly, for he had been made a partner of the firm, but neither was he a success. His colleagues believed that he had discovered a queer dimension between making and striking out and treated him with puzzled, deferential respect. One morning X informed those colleagues that Armbruster had requested the services of a Flyte team. The general effect was rather as if X had declared himself capable of flight, and had then actually leapt from the fourteenth floor and after extending his thin arms soared successfully toward the pale and yellow sun. That afternoon, the team – X, two Whippet-like partners, an associate with asthma who wheezed even when he walked, a computer specialist, X's private secretary, me – met Armbruster for perhaps five minutes. I had expected someone thuggish, but Armbruster turned out to be thin, even frail, with sandy-yellow hair, a long, sallow Talmudic face, concave cheeks, and a delicate arched nose flaring dramatically to a pair of midnight-blue and inky nostrils. He spoke with a trace of a lisp and a touch of a stutter. Standing in front of his desk, one lean hip braced against its edge, he outlined in delicate detail his dissatisfaction with the management methods of his predecessor and his vision of the wonderful world our two firms would enter together. "Suh, Suh, So much to do," he stammered softly, as X sat beaming, his own phocine head rocking in time to the barely pulsed beat of Armbruster's left foot.

I sat through the meeting in hyperborean gloom and took a full hour to return to my office, zig-zagging aimlessly up one avenue, peering peepishly into a store window and shrugging off the overtures of a shambling, purple-lipped, mid-afternoon streetwalker.

At the office, I threaded my way though a labyrinth of carpeted corridors until I reached the austere cubicle that I shared with another associate, a blonde engineer from Hamburg named Homburg.

"Haff you heard?" he asked, his accent thickening in anticipation.

"Heard what?"

"The Board just fired Armbruster."

"No!"

"*Ja, ja,* an hour ago."

X himself arrived at the office soon thereafter and

retired immediately to the washroom. From behind the closed doors he proceeded to emit a series of peculiar, agitated whoops. "What on earth?" said a secretary who had come to deliver a memorandum.

"He is womiting," said Homburg precisely.

Newton's Version

It is the aim of physics to limn the ultimate structure of reality by means of laws that are at once general and simple. This lapidary formulation raises precisely seven problems: What is the meaning of "limn," "ultimate," "structure," "reality," "laws," "general," and "simple"? I have no idea; and neither, I suspect, do the physicists.

The appeal of the Newtonian vision is much a matter, I think, of a glimpse glanced of a world that is clear in all of its aspects, unchanging, measured, determined, and regular. My own life may tremble precariously on the very cockroach cusp of chaos, but the Newtonian universe is bisected by "Absolute, True and Mathematical Time, which of itself, and from its own nature, flows equably without regard to anything external; [and] Absolute Space, which in its own nature, without regard to anything external, remains always similar and immovable". This is a formulation in which a man may find comfort without worrying overmuch about its coherence. Newtonian *mechanics*, as it has turned out, is incorrect in that bizarre limit in which objects contract, mass expands, and time itself slows and then stops. This circumstance is relevant only to those who demand of a theory that it be true. And besides, beyond Newton's version of mechanics, there is Newton's vision.

From one perspective, Newtonian mechanics is a theory of motion; from yet another, an elaborate exercise in the classification of certain geometrical figures. Under the spell of either incarnation, ordinary objects shed with fascinating speed their habitual properties of color, density, and texture: A billiard ball or a planet becomes a *point-mass*, the whole of its now perilous identity concentrated at its center. The Newtonian systems thus occupy rather a sparse conceptual stage. They tend to appeal to the man with a taste for desert landscapes.

The computer, I have observed, marks time in integral instants. Between the beats of its mechanical heart or clock, the machine has no life and slips into a temporary void. Newtonian time, by way of contrast, is continuous. Between any two temporal points, there is a third. These touchstones (or tombstones) go on forever. This image of Newtonian time may well suggest misleadingly that on this scheme time has a definite direction. Not so. The equations of motion that figure in Newtonian mechanics are *time-independent* (the physicist's term) or *autonomous* (the mathematician's); what has come and what is to come coincide, providing the physicist with his first, and only, view of the natural world under an aspect of eternity.

Wherever they are and whatever they are doing, the elements of a Newtonian system occupy some quite definite position and so establish themselves as a geometric *configuration*. The set of such configurations constitutes an abstract space of possibilities thickly enveloping any actual configuration of Newtonian elements in the here and now. Observing a particular Newtonian system – the solar system, for example – the physicist intervenes in its workings (intellectually if not literally) only to fix formally the system's *initial configuration;* he then aims to determine completely the future position of the particles and describe the trajectory of their motion. Doing the second, he has done the first.

On the level of metaphysics, it is apparent that things change only because something gets them to change. It is in the nature of physical curiosity to wish to make this declaration precise. Our own experiences on the surface of the earth suggest that everything that moves is moved by something, and requires thus a continual infusion of force if it is to remain in motion. Such was Aristotle's dictum – the doctrine of impetus. To the extent that it is true, it is plainly true *locally*, where baseballs and ballerinas rise in a graceful arc, their velocity diminishing in proportion to the force with which they were originally impressed, and the resistance that they encounter – the friction of the air, say.

There is something impossibly tender in the long tradition of Aristotelian mechanics. Absolutely nothing to which one can sensuously point indicates *obviously* that the Aristotelian view of things is incorrect. The discovery that it is occurs in the history of thought as a rude shock, and marks the point at which Nature and trust in the obvious collide, with trust in the obvious very much the bruised loser. The relationship between force and motion that Aristotle advanced, Newton rejected. It is only a change in velocity – *acceleration* – that is the result of an application of force – $F = ma$, to put the whole thing into four famous symbols. Where no force is needed, none need be postulated. In the absence of friction, an object moving in a straight line will continue to move in a straight line forever. In the absence of force, an object at rest – an undergraduate, say – will remain at rest forever.

●●●

As the physicist sees things, Newtonian mechanics is a branch of physics, and limited thus to the one (and only) physical world. The physicist is an intellectual Puritan in the narrowness of his affections. Yet there are as many mathematical worlds as there are consistent mathematical theories. The mathematician is an intellectual Mormon in the voluptuousness of *his* attachments and is forever appearing in public accompanied by ten lavishly complaining wives. Strangely enough, mathematics and physics are mutually absorbed; that these two separate disciplines should have anything to do with one another is a very great mystery.

Like any other science, mathematics begins with certain concepts too rich to admit of definition – the notion of a *function*, for example, something that serves to pair one group of objects with certain Significant Others. A function is thus a relationship between two or more mathematical objects. These need not be numbers. Unfortunately, the notion of a relationship is no clearer than the notion of a function. Relationships and functions may both be defined in the terms of set theory, a subject, as I have said, astonishingly rich in paradox and little else, but there the chain of definitions simply stops.

The simplest of all functions is also the most obvious and acts to pair a number with itself. Under its influence, the number 1 is mapped to the number 1, 2 to 2, and 999 to 999. The function acts as a mirror. In squaring a number, to consider another function, I pair a number to itself, multiplied just once by itself: 1 to 1, 2 to 4, 3 to 9. To throw the relationship into relief, mathematicians write $f(1) = 1$, $f(2) = 4$, and $f(3) = 9$. Here f is presumed to operate on 1, then 2, and then 3, and may be imagined as a kind of magician who, after draping parentheses around a given number, diverts the eye just slightly to the right, where a brand new number miraculously appears.

The process involved in pairing a number to itself may be expressed in its full and virile generality by means of algebraic variables – letters such as x, y, or w, which stand indiscriminately for any natural number: $f(\mathrm{x}) = x^2$, instead of $f(1) = 1$ or $f(2) = 4$. This is a handy form of shorthand, in which an operation of infinite scope is given by what amounts to a simple military command: Take any natural number, Buster, and square it.

•••

The drama of Newtonian mechanics plays within a space of three dimensions – an old standby from daily life – up and down, left and right, in and out; but however much we may wonder why our physical *experience* inevitably takes place in precisely three dimensions, there is nothing in the mechanics of the matter that makes three an especially important number. Within special and general relativity, the three dimensions of space jostle companionably for attention in the company of a fourth dimension marking time.

A straight line constitutes a space of but one mathematical dimension. Its analysis is complete when its origin has been specified, together with a unit of measurement. Imagine now two infinitely long straight lines crossed at the perpendicular. These are the axes of a two-dimensional *Cartesian coordinate system*. Each of those straight lines is numerically sub-divided. At their point of intersection, both are set to 0. The fact that these lines intersect makes it possible to represent negative as well as positive numbers in space. But a Cartesian coordinate system also depicts the plane, which is purely a geometric object, and so forges a connection between two parts of the mathematical experience – geometry and algebra. Each point in the plane, for example, may be uniquely identified by a pair of numbers – its coordinate – by the simple device of counting along each axis. With the addition of a third coordinate axis, a mathematical representation of space itself emerges. The specification of a point now requires three separate numbers. In elementary physics textbooks, triplets give way to *vectors*, a class of curiously rootless geometrical objects; but vectors and numbers admit of a mathematical merger or marriage. A two-dimensional vector is nothing more than a pair of numbers; a three-dimensional vector is a triplet; and an n-dimensional vector, n numbers frozen in a particular order.

Very simple mathematical ideas lie at the secret silent heart of physics. I shall elaborate. Newton thought to compare a particle to a mathematical point, and its trajectory – where it goes, where it has been – to a curve throughout the whole of space. This was an act of double abstraction and a queer initial point for a theory of motion: Points have no width or mass, and curves do not move. In any event, suppose that a particle passes through the origin of a two-dimensional Cartesian coordinate system bound for parts unknown. To the right is an axis marking time; above, an axis marking distance (but not direction). The speed of the particle is constant. The physicist achieves a *partial* physical description of the particle's behavior if he is able to say for a given time how far the particle has travelled from its origin; a *complete* description, if he is able to describe distance in terms of time. A relationship of this sort – distance against time – requires a function for its expression. And functions are *mathematical* objects.

The simplest (although not the most common) curve in space is a straight line. Strictly speaking, lines and curves are geometric objects and belong to a primitive intellectual kingdom. In analytic geometry, a subject no doubt remembered with a moan of pure misery by almost anyone past high school age, a lucid and invigorating connection is created between certain algebraic equations (the moan deepens) and a class of curves (the moan darkens), the effect as striking as that produced by an illustrated medieval manuscript. Pairs of distinct points in the plane (*four* numbers are required) determine a single straight line; there the thing hangs, dutifully extending itself forever in two directions. Straight lines may also be described algebraically by an equation: $y = mx + b$, for example, where b is the point at which the line crosses the y-axis, and m represents its *slope,* or angle of inclination. Four numbers are again required to express this equation; three to solve it.

In two dimensions, straight lines may ascend or descend, move vertically upward like a rocket, or remain horizontal like a thin stream crossing the plane from one remumbled infinity to another. The slope of a straight line, I have said, corresponds to its angle of inclination – the ratio of differences between successive values of y and x. A line pitched at a forty-five degree angle moves upward by one unit for every unit that it moves to the right. Its slope is 1. That plane, with its ascending straight line and crossed and martyred coordinate axes, the reader must now hold in suspension and flash-freeze as an image. A cyclist, imagine, dressed, perhaps, in those absurd and iridescent tights much in favor in California, is bent on moving from here to there. His motion I reflect by means of the mathematical mirror of a coordinate system – the very one the reader is now instructed to fetch from memory. Movement along the x-axis, as before, is movement in time; the y-axis measures distance. Units are fixed first in minutes and then in miles; the cyclist's behavior I reduce to, and then represent by, the trajectory of a curve. The routine ratio of distance over time now acquires a novel incarnation as *velocity,* a concept with no cognate in mathematics. Two minutes having elapsed, the cyclist has covered two miles. His rate of speed is precisely one mile a minute.

Only three physical concepts have made an appearance in this discussion: time, distance, and speed. How far a particle has gone is a matter of how fast it has been going, and how long it has been going fast. How fast a particle is going is a matter of how far it has gone, and how long it has taken to go far. Time and either speed or distance suffice to make the real world rise.

On the Arm

How about that? I loved working at Flyte and Company. And who wouldn't? Each and every morning, in fair weather or foul, I would breakfast at the Waldorf-Astoria, where the pancakes were very fine, taking the meal on the arm, of course, and listing it as product development. I would then amble over to our offices on Park Avenue. In the lobby of the building, an imitation of an imitation of the Lever House, and quite striking as a monstrosity, I would purchase the first of the day's cigars, generally a Don Diego or a Jamaican cigar, with an earth-clay bittersweet taste that I had come to favor. Then I would ascend majestically to the fourteenth floor and repair to my cubicle. My secretary, a very bouncy Irish girl with a snub nose, freckles, short legs, and sunny hips, would fetch coffee and the morning paper. Three days out of five I would take my lunch at the *Café Chauveron*. Often my brother-in-law, the Slug, who was professionally engaged in maneuvering his way into the hearts (and bankbooks) of several Park Avenue debutantes, would accompany me. Charles, the suave Belgian headwaiter, would greet us by name, a look of roguishness on his smooth, round, very poised face. Our meals, of course, were lavish affairs, with many separate courses and a good deal of wine. I generally charged the whole thing to executive recruitment, inventing for the Slug any number of distinct and separate identities, and encouraging him to avow, should anyone ask, that he was a Harvard MBA, or a Stanford systems theorist, or an expert on data management from Bombay in New York for a conference, or even, once, a Greek Orthodox priest much concerned with hydroelectrical power in the Third World.

The Vagaries of Curvature

Newton's law of motion associates force, mass, and acceleration. An object in uniform and rectilinear motion experiences no acceleration and moves across the coordinate plane (one axis representing time, the other distance) with smooth and untroubled confidence. It is thus, no doubt, that objects move in Hell. In determining the slope of a straight line, mathematicians take the tritest of ratios; because the line is straight, it does not matter *where* the slope is measured. Throughout the whole of space it stays stubbornly the same. Now, a straight line is the shortest distance between two points, but not the most common, as airline pilots and politicians are both aware. And, unlike a straight line, a curve is a creature of change. The equation $f(x) = x^2$ describes a parabola in the plane. At 1, $f(x)$ is 1, at 2, 4, and at 3, 9. The obvious procedure for fixing the slope of *this* line by taking ratios fails. Different points make for different ratios. But then again those physical objects that move without changing their motion are singular. Rising from the couch with a yawn and a stretch, *I* pass from a state of rest to a state of vigorous and untroubled activity. Acceleration. Reconsidering the whole business, I sink back onto the couch with a sigh and a moan. Deceleration. By means of a single superbly controlled effort (really rather like a panther, I have been told) I go quite beyond uniform and rectilinear motion. The equation that might describe what I have done yet again depicts a curve in space.

To the physicist, velocity appears as a *physical* concept arising spontaneously from purely a *mathematical* circle of definitions. So far, the mathematician has made himself clear only in the case of the straight line. The case of the common curve remains beyond a definition of velocity framed in terms of straight lines and their slopes. Still, some content, the physicist might urge, may yet be given to the *general* concept of velocity by averaging speeds.

The average velocity of a particle that has covered sixty miles in sixty minutes is just sixty miles an hour. This has all the appearances of a sane and sensible solution to the problem of acceleration. A particle in motion may, however, begin things in a state of rest and yet cover sixty miles in sixty minutes by means of a late, lavish burst of speed. On distributing the particle's average velocity to each of those sixty minutes, the physicist is apt to conclude that at any particular moment the damn thing is either moving faster or slower than it really is. This is an unhelpful position to have reached. What the physicist instead requires is the *instantaneous* velocity of a particle – some measure of its speed *now*.

In considering precisely this problem, Leibnitz and Newton thought to rescue the idea of a ratio. Not worried overmuch by logical scruples, they accepted without qualm the thesis that some numbers might be less than any other number and yet greater than zero. Such are the *infinitesimals*. The ratio of infinitesimal change in distance over infinitesimal change in time they reckoned the velocity of an object at a point. It was by this act that a suspension of mathematical disbelief was achieved. The great analytic mathematicians of the nineteenth century managed to eliminate this appeal to the logically unwholesome by means of the concept of a *limit* – one of the great, fabulous, fragrant flowers in the history of thought. The fractions $1/1$, $1/2$, $1/3$, when continued in the obvious way (the denominator getting larger), constitute an infinitely extended mathematical series. Somewhere beyond any of the fractions themselves, solitary, singular, seductive, is the limit toward which those numbers are patiently plodding – 0, in the present case. In the calculus, this elegant but general concept is specialized. At 2, $f(2)$ is 4. I retain the example of the parabola, but the action is now localized at a point. Suppose that h represents a small increment to 2 – 2 and a tad, so to speak. *This* increment is real and shares no disturbing kinship to those infinitesimals employed by Leibnitz and Newton. The ratio between $f(2 + h) - f(2)$ and h is a measure of the small degree in which changes in one direction along the axis of a curve are balanced by small changes in the other. Assigning this ratio to the curve as its slope, the mathematician remains in the domain of average quantities. Yet what happens when h *contracts*? Recall that f maps each number to its square. Hence $f(2 + h) - f(2) = (2 + h)(2 + h) - 4 = 4 + 4h + h^2 - 4$. When divided by h, this comes to $4 + h$. As h becomes smaller, the sum tends simply to 4. Tendings and tendencies suggest the existence of a limit. Precisely. A limit so defined is a *derivative* and functions purely as a mathematical measure of curvature; when reinterpreted physically, derivatives serve as stand-ins for the concept of instantaneous speed.

The calculation that I have just carried out may be extended so that each point on the curve is assigned a slope. In general terms, with variables standing in for particular numbers, f is $f(x) = x^2$. Consider now the behavior of $f(x + h) - f(x)$. This amounts to $(x + h)(x + h) = x^2 + 2xh + h^2 - x^2$. First and last figures in the numerator cancel; divide by h, and the result is $2x + h$. As h itself heads for home at zero, $2x + h$ reduces to the limit $2x$. The derivative of a function at a point is a number; the derivative of a function as a whole is yet again a *function*.

Just recently, I must mention, simply to complete the story, infinitesimals have made a reappearance, purged by contemporary logicians of their impurities. It is now possible to picture the slope of a curve in the old-fashioned way, a strange development in the history of mathematics in which a queer path is traced from the absurd to the austere and back again to the absurd.

Launched

Have I yet mentioned lunch? I am talking now of the ritual, and not the meal. On most days, associates took lunch with other associates and spent their time in bluff, hearty conversation marked chiefly by the ricocheting exchange of phrases – "Hey, Adcock, what's happening?" – whose length, I have since determined, are precisely calibrated to allow a man simultaneously to chew and speak. The partners, I suppose, ate with their clients, although I never saw any of them at the *Café Chauveron*. The directors lunched with one another at their private clubs. On joining the firm, every associate was taken by a director to his own club for lunch. My host on this ceremonial occasion was Ronald X, a tall, very quiet character, with a tendency to dress in suits cut for smaller men so that his suit jacket habitually rode over his rump; I had met him just once before during a painful interview in his office and had formed the impression that life at the top might possibly not be all that much fun. The club itself was somewhere in the east 80s; its name I have forgotten, but the club (or its name) had something to do with horses.

It was a fine, warm day in late June. Up the marble steps of the club we trooped, the four of us. Ronald X rang the buzzer that was marked "Members Only," and we passed instantaneously from the somewhat sordid New York streets into a silent sanctum, the passage and the transformation reminding me somehow of certain episodes on *The Twilight Zone*. It would be helpful were the reader now to hum that show's distinctive theme. A butler greeted us suavely, pampering our host, patronizing his guests. We passed through the reception room and entered the library for sherry. This was served by a bartender who had mastered the remarkable trick of seeming to bend as he served entirely from his pelvic girdle. My host endeavored to engage each of us in conversation.

When he heard that my own preparation for a career in management had consisted of the acquisition of a degree in analytic philosophy, he paused and allowed his eyebrows to commence a delicate waggling (the step before a frank furrow). "I have always believed," he finally said, "that what makes the firm so great is its ability to employ *diverse* talents."

Lunch was finally announced. We rose and heel-clicked our way over the corridors to the dining room. A good deal of time during the meal was given over to the discussion of the first course, the melon, which the club had specially grown in Maine. I pause to resurrect a memorable vision: Some mummy-necked Maine farmer patiently prodding the stone-hard ground with one stubby finger to uncover a single, pneumatic melon:

Ayuh, I grow them for these there imbeciles in New York, I do.

My host drew attention to the melon's texture, ripeness, and astonishing firmness: It was the only subject of conversation on which he was prepared to speak passionately. I ate with a fine concern that I might inadvertently shatter the solemnity of the occasion, hardly daring to chew my food. When another associate innocently praised the *canteloupe,* Ronald X raised his mobile eyebrows skyward in consternation.

"It is *not,*" he said frigidly, "a canteloupe."

The luncheon marked the last time any of us saw a director socially. Once in a great while, an associate, in a burst of bravado, might approach a director for lunch.

"Say Ron," he would say, regretting the whole business as soon as he caught the mackerel-chilly look in Ron X's cobalt eyes, "I was wondering if we might get together for lunch?"

A nod and a pause from cobalt-eyes.

"You're . . . ?"

"Adcock. Roger Adcock."

Another pause. Another nod.

"Ask Miss O'Dwyer to see what my book looks like."

Thus instructed, Adcock would slink over to Miss Dwyer's office, larger, in fact, than his own, and better appointed, and, by now thoroughly embarrassed, blurt out his criminal craving.

"Ron X asked me to check with you about lunch."

"I see," Miss O'Dwyer would say, in a tone of voice that suggested that Ron X might as well have asked to lunch with Martin Borman.

Miss O'Dwyer would then consult an elegant appointment book, turning the gilt-tipped pages with deliberation.

"He's not free this month," she would finally decide, the implication unsaid but plain that Ron X would *never* be free to lunch with this Adcock.

Zusammenhang

In social, political, and economic life, things tend to hang together in unsuspected ways – *Zusammenhang*. Elsewhere, they don't. A sense of the interconnectedness of things is the prerogative of the mystic and the mathematician. The mystic, of course, need only shut his eyes and touch together his thumb and forefinger to see everything come swarming up to consciousness, the essential Oneness too obvious even to describe. The mathematician attends to the connections among things by means of a system of equations in several algebraic unknowns. The concepts that he employs are not trivial. Even the inveterate mathematical illiterate, humming mantras or gingerly inserting his consciousness into the astral plane, cannot but thrill at the thought that they exist.

In elementary algebra, for example, one was forever being asked how many pounds of fertilizer the farmer could purchase if each pound cost twice as much as three pounds of tobacco, and tobacco is selling at fifteen cents a bushel. Questions of this sort appeared inevitably in a section of the dog-eared textbook entitled "Fun with Numbers." In my own case that farmer wound up purchasing several metric tons of fertilizer and paid billions for each bushel. Whatever fun there is comes later.

The algebraic equations of the eighth grade, and grades beyond, explain how it is that some things are connected; the analysis is static. Algebra makes no provision for change. In the elementary calculus the mathematician starts with a function and ends with its derivative. Now a function, I have said, is a relationship or rule pairing one quantity with another. In the calculus the initial function is *known* and enters into all subsequent calculations with a self-satisfied sense of its own identity. The problem and procedure of the calculus may be reversed.

Imagine that a clutch of rabbits are, as rabbits will, copulating industriously, and thus augmenting the stock of rabbits available for fricassees or fondling. The time is the present. There is no way to make the example more interesting. Prompted chiefly by the desire to make a pedagogical point, the mathematician asks how many rabbits he might expect to have at some specified time in the future. Even without any mathematics at all, it is plain that the solution to the problem resolves itself broadly into two components: The *speed* at which the rabbits are mounting and mating one another, and the *number* of rabbits so engaged. Speed suggests a derivative, and the varying numbers of rabbits, a function mapping time to rabbits. These items are obviously related. For the moment, the function – $f(t)$ – enters into all further calculations with its identity opaque. What of speed? The derivative of a function expresses a ratio taken at the limit; in the case of rabbit reproduction, the extent of this ratio depends on the number of rabbits and is thus itself contingent upon $f(t)$. Contingent how? Contingent thus: At any moment, only some constant percentage P of rabbits are reproducing themselves. Sexual speed depends on P and $f(t)$ itself.

An equation is a mathematical enterprise in which an identity is expressed. One half of the identity – the derivative – is in place. The other half now admits of expression as the product of P and $f(t)$. The result is a *differential* equation: $D[f(t)] = P \cdot f(t)$. From the equation, a simple English declaration may be recovered: The instantaneous speed at which rabbits are reproducing is given as the product of the number of rabbits and a constant.

A single differential equation contains but one unknown function, and resembles thus the very simplest of algebraic equations – $4x = x + 3$, for example. This is fine, as far as it goes, but often the mathematician is interested in tracing the dynamic relationship among several changing processes or planets. In elementary algebra, this need is satisfied by a *system* of algebraic equations in several unknowns – x and y, for example, as in $5x = y + 3$, $x + y = 10$. These equations must be solved jointly. The conditions expressed in each are contingent for their resolution on the conditions expressed in the other. When it comes to differential equations, unknown functions replace unknown algebraic variables; but the underlying idea, with its mysterious suggestion that solving a system of equations is rather like keeping in fixed focus two dissimilar phases of the same picture (the wine glass *and* the woman) – this remains precisely the same.

A Mozart of Management

I had spent a few weeks on the corporate side of things; now I was engaged in drawing up a financial plan for the New York Health and Hospitals Corporation. I knew nothing of finance and little of health. My task was to prepare a multi-colored chart indicating the flow and ownership of various public funds. Those points of factual detail I found troubling I fabricated. My superior in this effort was Carter Z, the firm's youngest partner, and a great Mozart of management. It was Carter who had initially recruited me to the firm during an interview made notable by the degree to which each of us said only what we expected that the other wished to hear. For my part, I had formed the lunatic impression that in its public practice the firm was interested in improving the delivery of city services, and so suggested during every pause in the conversation that I had something more in mind in working for the firm than taking my meals on the arm or making a boodle. For his part, Carter had formed the equally lunatic impression that my chief reason for leaving the academic world was an insufficiency in the opportunities it afforded me to serve my fellow man. On those occasions during our conversation when I was not expressing my interest in good works, Carter was. Thrusting his great globe of a head toward me, his blue eyes bright, and resting his forearms on his own knees, he said: "Not much time left. We've got to act *now*!" With that said he sat upright and punched his palm with his fist.

Not knowing for the first time just what to say, I said that *absolutely*, we had better act now.

Celestial Dynamics

Force impresses itself on matter by acceleration. This marks the difference from moment to moment in an object's speed. The velocity of any object is its change in position against its change in time. Acceleration is thus the derivative of velocity. Newton's second law of motion states that force is given as the product of mass and acceleration. Knowing the forces that impinge upon an object, the physicist may calculate its acceleration, and with its acceleration, its velocity, and ultimately, by means of integration (another radiant, multi-hued mathematical concept – the cousin and correlate to differentiation), its change in position.

All of this lies somehow on that tentative border between mathematics and physics, and offers a contrast between two sets of concepts: the delicate Greek tangle of the calculus, the simple Roman brutality of Newton's law.

It was Newton's vast ambition to provide a complete analysis of the mechanical behavior of a system of interacting particles. What lent to his work its imperial boldness was his insistence that one and the same analysis might apply indifferently to *any* system of interacting particles. This served to unify in one mathematical figure the behavior of objects in motion on the surface of the earth and the trajectories of the planets themselves, two systems that no one had thought to bring under the impress of a common mathematical description. In a single stunning year, Newton, working alone, and absent from London because of the plague, described and then solved the equations of motion for an interacting system of mechanical particles, and invented the mathematical tools, chiefly the calculus, that were necessary to articulate the whole of his vision.

By means of the peculiar and unrepeatable alchemy of genius, Newton solved completely the abstract problem of describing the universe as a mechanical system, at least in the sense that he was able to state for an indefinite number of particles or planets the laws governing their evolution. Within celestial dynamics, that most musical of sciences, the solar system is treated as a Newtonian mechanical system. In the century before Newton's birth, Johannes Kepler had observed that the orbit of those planets that he could see described an ellipse. Kepler's law, Newton knew, sufficed to calculate the acceleration with which the planets revolved around the sun, and with their acceleration, their trajectory. It was in the course of carrying out this calculation that Newton was led to the inverse-square law. All objects in the universe attract one another, Newton discovered (or decided – the line between fact and fabrication is very thin in physics), with a force inversely proportional to the square of the distance between them, and proportional to their mass. The universal force of gravitation, although affected *by* distance, acts *at* a distance, through empty space and across the inhospitable regions of the stellar night. The inverse-square law holds not only for the sun and its satellites, but for all material bodies wherever they might be found, in the observable regions of the solar system or in the depth of space, where unimaginable cold predominates. By treating the solar system as if it consisted entirely of two bodies – the sun and a planet – Newton was able to demonstrate that to the extent that these bodies were attracting one another in accordance with the inverse-square law, the planet, in tracing a curve through the sky, would describe an ellipse. He thus returned to precisely the intellectual point from which he had departed. There is a very pleasant symmetry at work in all this, an engaging play between fact and theory, with Kepler and Newton, resurrected now in a timeless world, pointing weakly at each other (crooked fingers, faint smiles), like athletes too pooped to do much more than pant.

Within a Newtonian universe change comes about when particles alter their position in space. It is thus that the chugging steam train moves across the landscape of elementary texts, black smoke billowing. From a more abstract point of view, in which there is less to see and more to appreciate, those changes undergone by a Newtonian system involve a systematic process in which *states* of the system instantaneously give way to other states, as when with a certain pressure of thumb and forefinger, the magician gets the cards of a deck to snap forward in a series of pale, repetitive scraps. Whatever intuitive meaning Newton's laws of motion may have – this usually illustrated by ricocheting bullets or billiard balls – they function mathematically as *mechanical* principles and serve in the history of thought to endow the concept of a machine with almost all of the content it ever acquires.

And here is the point to a puzzle. A mechanical system, we are inclined lightheartedly to say, is one that enjoys an unusual analytic transparency. In large measure, we understand the very concept of determinism – as in *I hadda do it* (usually an assassination or assignation) – only by reference to Newtonian mechanics, and not the other way around. What the system does is determined by what it did. Knowing the latter, we know the former. *Pas tout a fait*. The great majority of differential equations do not lend themselves to analytic solution, and lie there obdurately on the page, the precise and global nature of their destinies inscrutable. In systems comprising two bodies Newton's equations of motion admit of a complete and closed solution. When three particles interact – only one more particle, after all – the Newtonian system that results cannot be solved. Three is the number of the Trinity, and the number, too, at which the universe ceases to be computable – evidence, if any were needed, that in science, as in the rest of life, one is dealing with a form of irony.

Happy Days at HRA

By and by, I found myself working as a senior quantitative analyst for the Human Resources Administration. Every morning I would insert myself manfully into the steaming subway cars of the IRT, and stand for an hour or so packed into a stenchy, wenchy mob, until, with a groan and a shudder, the train would stop at Church Street, and I would elbow and shove and push my way toward the door, my progress slowed each day by the very same sober-eyed, unbudgeable fat man, or his brother or his cousin, their arms (thinking of them now as an entire family of impediments) locked around the subway car's badly placed central post, a copy of the *Daily News*, which they read while moving their six lips slowly, folded into sections in front of their vigorously Blue-Cheesed noses.

At the HRA, I was asked to assist the deputy director of technical services, a frazzled, manic, frizzy-haired technician whose name I seem to have forgotten. Distracted to delirium by his own incompetence (and my own, no doubt) he lurched through the days that I spent in his company as if he were being erratically charged by some direct and free-floating current of electricity to which he alone was sensitive. The main-frame computers in the department's computing facility were in the hands of a crew of city civil servants: black women, mostly, with fierce eyes peering out bleakly from their fat and angry faces. Their attitude toward the machines they were operating was one of dark and sullen contempt. When the system went down – something that happened often – the women would leave their consoles in a group. "System down," they sang out happily, rising from their consoles in a heavy-haunched wave. Poor Drecker – that was his name! It comes back to me now. (Hi, Drecker. I hope you're having better luck these days.) At the end of my brief stay there I asked him to diagnose the disaster in which he was embedded. "Fuckin' A," he said, holding his twitching feet and fingers still for a brief moment and rolling his eyes upward in a gesture of resignation and despair.

"Fuckin' A."

Newton's Vision

At a point roughly midway between Richard Nixon's first inauguration and his hysterical and hilarious farewell speech, it became fashionable among writers and thinkers (a nice distinction, that) to see any number of ecological or social catastrophes coming in the short term. Jay Forrester and Denis Meadows, for example, studied models of the world's economies and reported themselves convinced that disaster was only decades away; at the *International Institute for Applied Systems Analysis* in Vienna, where I spent a wonderfully beery three months, great global models depicting the production and consumption of energy were created on the computer, with results that suggested that someone had better do something promptly. During the last years of President Carter's luckless administration, systems analysts constructed a global model of the world so bleak in its conclusions as to confirm the President in his conviction of the ubiquity of sin and the inevitability of its punishment.

It goes without saying, of course, that since time immemorial someone somewhere has stood solemnly in some marketplace or other, warning the rest of us that we had better get our acts together: Don't smoke, don't drink, don't do drugs, don't despair – a quartet of remonstrations obviously impossible simultaneously to satisfy. There is an interesting difference, however, between then and now. The Hebrew Prophets were much concerned with sounding an alarm among the Hebrew people. What happened to primitives elsewhere was not their concern. This has always struck me as a sensible attitude. The first pictures of the earth from outer space, however, suggested to almost everyone that as far as life goes we are all in this together. Contemporary concerns are thus *global* in that trite and tenuous way in which it is often said that the world has become a global village (Redwood City and not Florence, unfortunately). No doubt, the idea had occurred before: Those pictures gave to the idea a dramatic charge. But the process by which a leading theme is translated into a living theory is one crucially contingent on the availability of certain concepts. The business of seeing the world as a single object, in which a great many diverse things and processes hang together, requires some tolerable notion of a *system*. This idea came late to human consciousness, and expresses, if anything, the world view projected so patiently by Newtonian mechanics – Newton's vision as opposed to Newton's version.

No less than any other theory, Newtonian mechanics achieves its singular force by means of a concentration of its intellectual resources. A natural law – $F = ma$, for example, or the inverse-square law – represents the intense, bittersweet liquor that remains when a concept is steadily distilled. The physical world is in its largest aspect informed by the concept of *force*; the rest of the universe, (where life, love, luck and language hold sway), by the concept of *growth*. Within Newtonian mechanics, force is ultimately explained by the laws of force, but in biology, or the social sciences, growth is tied to no correlative natural laws. The very simplest equations of growth, for example, suggest that at any given time an object, or collection of objects, grows in a way that is proportional to its size. When expressed precisely, models of growth admit of a simple solution: Things grow exponentially. This way of looking at things is hardly an improvement over the declaration that often things grow, generally until they stop. The mathematics is ceremonial.

Newton's vision, I have suggested, is much a matter of the projection of Newton's version of mechanics onto an alien screen. What has stayed in focus under this projection is simply the concept of a system. Is there any reason to suppose that anything else will ever come clear? I am inclined to answer my own question with a laconic *nope*; but making this point plain would require a separate argument, another book, another life.

Just Like That

Aside from waking up – nothing to be sneezed at, of course – the chief pleasure of the morning is the daily newspaper. Turning to my own, I discover, among other things, that General Zia and the American Ambassador to Pakistan have just perished in a plane crash. (Gee fellahs, that is bad luck. Just as you made it to the International Big Time, too.) This pleasant and dramatic bit of news prompts me to a minor mathematical meditation. Day by day, I suppose, we are alive, until, one day, we are not. In life, in love, in politics, and in art – but *not* in the calculus or the theory of ordinary differential equations – things happen suddenly. The balloon that is blown up bursts. So do love affairs (*You did what? With whom?*), and reputations (*Don who?*). Voices and nerves suddenly crack; soufflés and airplanes fall at the worst possible times. Quite out of the blue, your neighbor's pit bull sinks his fangs into the mailman's tempting thigh, leaving your neighbor the unenviable task of explaining to an indifferent world how it is that Rover is generally the most reliable of animals, a perfect pussycat, in fact. Like Great Britain and Argentina, or Libya and Chad, or Iran and Iraq, nations go to war overnight. The ski or the worm turns; infatuation turns promptly to irritation. Clouds loom up in the night sky and dump their loads without warning. An artery ruptures during intercourse (Honey? *Honey!*). The scalpel slips in surgery (*Damn!*). Teeth suddenly break in mid meal. The roving hand, like the roving heart, encounters a sudden spurt of steadfast resistance (*Not now, Henry, not there!*). The heart stops beating.

Just like that.

The Right Stuff

Despite many claims to the contrary, modern mathematics remains a subject of interest chiefly to specialists; even physicists tend to reject, say, monodromy groups or algebraic K-theory as items that are altogether too much like Parcheesi in their combination of inaccessibility and frivolousness. For their part, mathematicians are not entirely unhappy with the thought that absolutely no one knows much of what they do; when pressed to explain themselves in public on even the most transparent of mathematical topics, they tend to fidget and mumble and ultimately confess that the thing is simply beyond them. The emergence of *catastrophe* theory has offered an interesting exception to this pattern.

Like a mosquito, catastrophe theory came to creation in stages. Early on, the French mathematician René Thom had taken to loitering among the speculative biologists, a group traditionally gifted neither in speculation nor in biology. His own credentials as a mathematician were exquisite; he was thus able to keep the company of his new colleagues with dazzling intellectual assurance. From the first, he flabbergasted them by declaring that their concrete concerns, whether in embryology, morphology, or molecular biology, were incompletely expressed without such stuff as differentiable germs and stratified jets. The British biologist C. H. Waddington introduced Thom's ideas to a larger world. Sceptics observed Thom's prominent walk and waddle to the cutting edge of theory and happily recalled the dismal experiences of mathematicians in biology. Before the cracking of the genetic code by purely biochemical means, any number of capable mathematicians had glutted the journals with code-theoretic conjectures that in retrospect resolve themselves into those that are wrong and those that are wrong again. Thom himself urged his views with majestic contempt for those scientists whose training did not include the study of smooth maps. "We have to let biologists busy themselves," he would later write, "with their concrete – but almost meaningless – experiments; in developmental biology, how could they hope to solve a problem they cannot even formulate?"

In 1972, Thom declared himself for mathematics at book length. Biologists reviewing *Structural Stability and Morphogenesis* reported themselves impressed with Thom's unwillingness to allow his ignorance to impede his research. Still, the usual fashion among mathematicians writing in areas outside pure mathematics has been to alternate between lunatic certainty and inane equivocation. Whatever its flaws, *Structural Stability and Morphogenesis* is a mature work of the imagination. Its very existence, therefore, struck most mathematicians as a miracle.

Christopher Zeeman, for example, read Thom with positively Homeric enthusiasm. In a series of widely read papers, touching on biology, sociology, prison reform, animal aggression, and political science, he championed the new theory, claiming for it a status as a mathematical novelty roughly comparable to that achieved by the calculus. Zeeman was himself an influential and distinguished topologist. His voice carried conviction, and it was heard. Rumpled and bearded, Zeeman appeared before audiences with an almost ancestral fear of fancy mathematics; with a few papers clutched in his bunched fingers, he would stand before the blackboard, rocking like one of those mechanical birds that perch on water glasses, and lay out the foundations of catastrophe theory with such surefooted charm that even sceptics in the audience would perk up; before mathematically minded audiences, he was simply incandescent.

No one, it goes without saying, without a professional concern for the singularities of smooth maps actually read Thom's book. The material that it presents is incomprehensible to the layman and out of the reach of even ordinary mathematicians. The book's natural antagonists suspected that were they to avow that *Structural Stability and Morphogenesis* contained nothing more than a string of solemn absurdities, the inevitable result would be an embarrassing exposure of their technical incompetence.

It thus remained for specialized mathematicians to address the theory critically. Early reviewers – John Guckenheimer, for example – had already observed, with just a touch of asperity, that Thom's program in pure mathematics was by no means free of difficulties. Hector Sussmann and Raphael Zahler, in their turn, considered the application of catastrophe theory to biology and the social sciences. "No CT [Catastrophe Theory] model that we have seen," they wrote, "is quantitatively correct, and the qualitative conclusions drawn are frequently wrong or vague or tautologous." Of course, Sussmann and Zahler took every precaution not to attack Thom directly on mathematical grounds. They are sensitive to the distinction between critical vigor and sheer foolhardiness. Yet they managed to suggest that for all his undisputed greatness as a mathematician, Thom was nonetheless befuddled – a type distinctly French in his combination of studied obscurity and inconclusive prolixity.

Their criticisms provoked the most glorious of controversies within applied mathematics, a field not known for the flamboyance of its exchanges. Zeeman's defenders have been somewhat slow in coming to his defense, and not being used to the high standards of indignation common in, say, comparative literature or linguistics, they have found themselves without an effective vocabulary of objurgation. The best that they can muster is the claim that Sussmann an Zahler indulged in misquotation and misrepresentation. Guckenheimer went so far as to remark in a letter to *Nature* that Sussmann and Zahler were "snide" – a gesture likely to be about as useful in confronting their raucous jeers as that of a peacock spreading its tail feathers before a boa constrictor.

Most mathematicians, I daresay, were secretly pleased by the unexpected turn of events. Those who knew nothing of differential topology saw in the controversy an excuse for not learning more; differential topologists who had stood by the sidelines as Zeeman capered in the limelight felt encouraged by Sussmann and Zahler and, looking backward, saw in their own silence only evidence of a strong and savvy sense of self-preservation. Biologists are always happy to be confirmed in the hope that organic chemistry is the last difficult subject that they will have to master. Only the social scientists seemed genuinely saddened to hear the shouts. To know a little differential topology gives a man an inestimable cachet among his peers in sociology. The social scientist who could actually distinguish between a torus and a donut (only the donut can be eaten) felt correctly that the polemics threatened to rob him of a hard won point of status.

I take absolutely no overall position. My interest in catastrophe theory and its critics is purely *pour le sport*.

Differential Topology

In geometry, things are what they are; within topology, things bend, stretch, contract, twist themselves indecently, and generally regard any shape into which they may be continuously deformed as a legitimate extension of their own identity. In *differential* topology, as in differential geometry, the methods of the calculus are brought into play. Functions of several variables are treated, of course; the manifolds over which they range are higher-dimensional analogues to the ordinary sheet or the differentiable curve traced over the plane by a smoothly moving pencil. Such structures have strengths that are strictly local. For any point on a manifold of dimension n, there exists a neighborhood of that point homeomorphic to the interior of an n-dimensional Euclidean sphere. On a manifold, things are seen as Euclid saw them. Between manifolds continuous mappings snake from one surface to touch lightly on another. If a mapping and its inverse are both differentiable, one has a *diffeomorphism*. Manifolds and diffeomorphisms are what one gets in differential topology.

This is a thumbnail sketch. Like any mathematical theory, differential topology comes into precise focus only when looked at from short distances – nothing more than nose length. Still, my reader should see that differential topology makes contact with the theory of ordinary differential equations, a subject with no focus at all, and an unattractive tendency to lurch and stagger over the entire mathematical stage.

In one sense, the theory of ordinary differential equations shares with elementary algebra an unwholesome fascination for the unknown. Given $D[f(x)] = F(f(x), t)$, what function, if any, can go proxy for $f(x)$? Who knows? Nineteenth-century analysts proved that unique solutions to such equations exist, but there is generally no getting at them analytically, and even now the mathematician who takes it into his head to solve a differential equation must make use of some scheme for numerical integration.

But in another sense, ordinary differential equations are geometric rather than analytic objects, and exhibit as geometric objects a particularly satisfying palpability. Take as a *state* of a system of ordinary differential equations $D[f_i(x)] = F_i(t, f_i(x))$, $i = 1, 2, \ldots, n$, the smallest set of numbers needed at t to predict the course of the system at points beyond. When i is n, the size of this set is fixed at $n + 1$. Such sets, taken together, constitute a *state space* – $\mathbb{R}^{n+1}$ in the present case, and an old standby from courses in calculus, where it puts in a pedagogical appearance as an $(n + 1)$-dimensional Euclidean vector space. Now F, understood as a function, acts on $\mathbb{R}^{n+1}$ to pair points in space and vectors. The result is a *vector field;* when a particular state space and its associated vector field are considered jointly, the geometric eye merges them into a single *dynamical system.*

Classical physicists studied such systems in the hope of solving their associated differential equations, an effort that in retrospect seems rather like using a screwdriver to hammer nails. Knowing one solution to a set of ordinary differential equations, the mathematician yet knew next to nothing about the others; more often than not, with the exception of certain celebrated successes that wink luridly from the literature, the effort involved in working out even a single satisfying solution to a complicated system of equations is disagreeable.

Stumped by just such troubles in celestial mechanics, Poincaré made the enchanting discovery that it was possible to study dynamical systems without solving those grim equations to which they were affixed. Complex and athletic computations of the sort that Jacobi had flamboyantly mastered, Poincaré ignored. Engineers and physicists might need specific solutions to special equations, but the mathematician with a feel for the subject's sure center could skip such scruffiness altogether. What counts is the global character of all solutions to a set of differential equations. By "global" Poincaré meant the behavior of the solution set as time stretches to infinity, and by "all solutions" he meant, quite simply, the whole kit and caboodle of them. Such are the *flows* to a differential equation. A lesser intelligence might well have drawn the obvious inference that in this way lies madness. Even now it seems a piece of bluff and lunatic optimism to assume that something sensible can be said of all solutions to a set of differential equations when no one solution is actually worked out. But Poincaré saw that the solutions are shaped by the geometry of the state space and the analytical character of the vector field. These come first. Knowing this, the mathematician can see, in his mind's eye, at least, simple solutions to sets of differential equations streak out over the plane with all the inevitability of iron filings being lined up by a powerful magnetic field. Nor need it be the plane that solution curves fill. Wrap the plane around a rod, and you get a cylinder; bend the cylinder so that its ends meet and you have a torus. Differential equations can run on such surfaces; for that matter they can run on surfaces as abstract as a differential manifold. But then the study of flows and the study of diffeomorphisms turn out to be tolerably close, one to the other, since flows are nothing more than the diffeomorphisms of an abstract manifold onto itself.

The qualitative or geometrical theory of ordinary differential equations was Poincaré's invention, but in looking over what I have just set down, I see that I have suggested that somehow Poincaré saw in a silent flash the whole interpenetration of differential topology and the theory of ordinary differential equations. Not so. Poincaré had conceived of differential equations in terms of flows or trajectories, and this inevitably served to highlight an aspect of the subject that had long lain in shadow. The Dialectic, moving calmly through the agency of such mathematicians as Birkhoff, Andronov and Pontryagin, Morse, Whitney, Peixoto, Milnor, Thom himself, Smale, Kupka, Williams, Ansonov, and Arnol'd, required yet again almost three quarters of a century to achieve a consolidation of these ideas.

Soli Deo Gloria.

The Triple Concepts

Much like a teenager recumbent on a couch, catastrophe theory sprawls over much of mathematics with a fine disregard for any natural contours. In its essentials, Thom's *theorem* represents an episode in the development of a single subject – the theory of the singularities of smooth maps; beyond the essentials, Thom's *theory* drapes and droops over almost every other part of differential topology.

First things first. A function, I have said, is something like a rule or regularity pairing one item with another. In the elementary calculus, functions of one real variable are studied; in differential topology, the functions play over many real (or complex) variables. Playing thus, they are renamed *maps*. A map pairing points in one plane with points in another sends *two* numbers to *two* numbers. The smooth maps are those whose derivatives are continuous. Smooth mappings between manifolds make up the most general context afforded by the theory of smooth maps.

By temperament, the physicist is interested in tracing causes to their effects. On being struck by an apple, Newton wondered why the thing fell toward the earth. It is a question and a concern that no mathematician would find rewarding. In treating the space of smooth mappings between manifolds, his hygienic ambition is to show where things belong and how they hang together. Counting, after all, is a form of classification, and classification nothing more than an abstract form of counting – in either case the first, and often the last, step in bringing order out of chaos. But the mathematician hankering for order *here* is bound to contemplate a space of this size with the glaze of disappointment over his eyes. Local pathologies are everywhere; where there is no pathology, there is rebarbative complexity instead.

To the philosopher, and I write as one, the grossness of complexity (or the complexity of grossness – either way) prompts a helpful hint: Leave something out. This is good advice – the philosopher's speciality. But what?

Confronting just this question in other contexts (the

theory of semi-groups, Lie Algebras, Knot Theory), the sophisticated mathematician very often has recourse to a single all-purpose consideration – *stability*. Now, in mathematics, stability is a concept that slides up and down a scale of specialization, which, in differential topology, at least, ranges in pitch from structural to infinitesimal stability. The simplest definition is purely topological. The mathematician, let us suppose, is presented with a set of X of objects; he is able to say – the suppositions are continuing – what it means for members of the set to be relatively *near* one another. And he is prepared to specify the conditions under which two objects in the set are qualitatively the same. In coming to a classification of X, the mathematician counts it as a blessing if he need only deal with those objects x such that every y in X that is near x is qualitatively the same as x. These are the stable objects; the property that they share is *environmental*, and obviously useful to the extent that it allows the mathematician to dismiss the differences among objects.

In fashioning the concept of a *well-posed problem* in analysis, the French mathematician Jacques Hadamard had in mind some principle of selection that might mark those mathematical systems with physical significance. His own list contained three points: Solutions to a well-posed system (of equations) must exist, they must be unique, and they must vary continuously with variations in the initial data. This last condition arises as a concession to the frailties of human measurement. "Data in nature," Hadamard writes, "cannot possibly be conceived as fixed; the mere process of measuring them involves small errors." Now theoreticians of quantum mechanics have stressed that to measure is to distort, but there is no need to repair to the formalism of quantum mechanics to elicit this observation. No form of measurement is quite without error, if only because all observations are finite. But observational error – the gap between what is and what is recorded – appears in the record of any experiment as a *perturbation*. And processes that are resolutely immune to perturbation are stable. Another sense of stability has now been introduced.

This is to describe things in the abstract. I am interested in mappings and not arbitrary objects; those mappings that are of special importance are again distinguished by their differential structure – the set of all C^k mappings between M and N, for example, where M and N are differential manifolds, and C^k indicates continuous differentiability up to order k (the derivative of the function, thought of as a function, has a derivative, and so does *its* derivative, the process iterated k times). To give some sense here to the notion of an environmentally stable mapping it is enough, generally, to specify a suitable measure of equivalence. *Aussitôt dit, aussitôt fait*. Two mappings, f and g, between manifolds are equivalent if the first may be changed to the second by means of a diffeomorphism, and vice versa. A definition of stability follows at once.

Once given, however, the definition specifies only the global properties of a stable mapping. This is a concept, like nuclear deterrence, too strong to do much good. A function may well be stable at a point and yet not stable altogether. This reservation prompts a definition of *local* stability, or stability at a point. Since the definition is local, it may be given in terms of Euclidean spaces instead of manifolds. By a *perturbation* of a function, I mean just what the physicist means – a conceptual tap of the sort that sends $f = x^3$, for example, to $g = x^3 + ux$, where u is small. A function f is locally stable at a point p if diffeomorphisms send f to g and g to f. A locally stable f stays pretty much the same even though the thing is bent slightly out of shape.

In the definition of environmental stability, a single, multi-faceted concept is resolved into the coordinating concepts of nearness and similarity; in the definition of local stability, the resolution is in favor of perturbation and persistence. This makes for four concepts, two definitions, but only one elusive idea.

The philosopher, W. V. O. Quine has observed, should tolerate no entity without some specification of its identity. From this counsel, the mathematician derives a corollary: There is no specifying the identity of an object without some assurances as to its stability. Whatever the general meaning of this declaration, the concept of stability appears in *this* discussion with its importance ratified by both the mathematician and the physicist. I have brought in the philosopher just for the fun of it.

•••

As an enterprise, classification is contingent upon the stability of the objects being classified – their willingness to sustain a qualitatively measurable sense of themselves in the face of perturbations. Yet the stable objects of a mathematical system may be in short supply, and the classification that results irrelevant because it involves the organization of a minority of mathematical objects.

In a great many very ordinary circumstances, what is at first singular gets explained by an appeal to what is typical. It is thus that we generally answer *Why'd He?* with *Wha'd you expect?*, an exchange that when generalized suggests that among other things science involves a search for the obvious. Within mathematics, a single all-purpose concept – *genericity* – stands in for the brute notion of being typical. Let X, as before, be a set upon which a suitable topology has been defined. A subset A of X is *open* if every x in A is contained in an open ball contained again in A, *dense*, if every point of X is arbitrarily close to points of A. A property P of elements of X is *generic* if the set that satisfies P in X contains a subset

A representable as a countable intersection of dense and open sets.

The generic objects in a mathematical world constitute a class that is quite large; its complement, correspondingly, is small, and, when measures are available, closes in on Measure 0, the mark of the mathematical miniature. Openness and density, taken jointly, comprise a property with two remarkable features. If A satisfies some generic property P in X, then *any* x in X may be approximated by some y in A. Generic sets are dense. But they are also open, so *no* x in A may be approximated by a y in X^c, where X^c represents those elements in X but not in A.

The discovery that sets satisfying P satisfy P generically is often all that the mathematician requires in order to complete a classification. Asking himself *why* x is P, he is now able to answer that *most* x's are P. If x exhibits P then so does any y sufficiently near x; if not, then some y that does can stand in for x to any degree of fineness. In the theory of linear ordinary differential equations, for example, the operators that generate hyperbolic flows form a set in the space of all operators that is both open and dense. This gives a very congenial structure to the space of such operators. In the space of smooth maps, stability already figures as an important concept. P might thus be identified with either environmental or local stability.

Under this identification, the classification problem for the space of smooth maps acquires a satiny pliancy. Counting in this context means something like reckoning with the stable maps. The exhibition and analysis of a generic set of stable maps empties the classification problem of its interest; beyond this, there is nothing much left to discover. Of course, one can have stability without genericity, and vice versa. The structurally stable vector fields are not generic in higher dimensions; the rational numbers are not stable under the operation of taking square roots. But from time to time, as in Morse theory, one has both genericity and stability. The effect is altogether gratifying.

•••

In its most elegant aspect, the classical calculus traces the interplay between the analysis of a curve and the characterization of its derivative. In the plane, a function $f(x)$ leaves a record of itself in the form of its *graph* – the curve representing its values. The singular points to a function mark the spot where its associated curve turns up or down. There, the derivative of f goes to 0. If f is singular at x, x is a *singular point* of f; $f(x)$, its *singular value*. Points that are not singular are *regular*; so are their images. This makes in all for a simple two-part classification. But the information afforded by this system is generally of little use. From the behavior of the first derivative there is no telling whether the curve is at a maximum, a minimum, or whether, perhaps, it is simply wandering about a point of inflection. The fact that the derivative is 0 means only that f is neither an increasing nor a decreasing monotonic function.

A much finer classification of curves comes about under the action of the *second* derivative of f taken at a singular point. The connection is actually gloriously simple: If the second derivative of f is greater than 0 at x, f is concave; if it is less than 0, convex. If the second derivative of the function is itself 0, a certain opalescence once again covers the scene. For the moment I describe this particular color as the mathematician might – the point so infected is *degenerate*.

The connection between a function and its derivatives is simple but still subtle. With the introduction of the second derivative of a function, the system by which its points are classified doubles in scope. There are the regular points and those that are singular; the behavior of the second derivative then splits the singular points three ways. This makes for a space of five possibilities. Correspondingly, a five-fold classification of curves in $\mathbb{R}^2$ segregates those that are monotonically increasing or decreasing about a given point and those that are not, and then those that are concave or convex. This makes for only four types: There is yet that nacreous degeneracy. But however the curves are counted, the extraordinary point remains that my five-fold classification of singular and regular points, which depends only on the exquisitely local properties of the first and second derivative, *induces* a corresponding classification on maps. (When I talk about an extraordinary point I have in mind a delicate double vision in which two mathematical epiphanies are superimposed. The first occurs as the imagination records with a rush the movement from analysis to geometry; the second, experienced simultaneously, but with a larger and more leisurely wave of appreciative amplitude, as the imagination begins to see in the fine structure of analytic objects – points and their properties – the tantalizing traces of a larger and more spacious geometrical world.)

In higher dimensions functions give way to maps; the derivative itself reappears as a linear map, a sort of souped-up straight line. Although the derivative of a map is yet again a map, it is very often convenient to represent the effect of the mapping by means of a *matrix* – a rectangular array of numbers. It is the *Jacobian* that stands in for the first derivative of a given map; the *Hessian*, for its second derivative. I spare the reader the details.

What of the singularities? In the plane no one can miss the spot where, like peaches suspended in Jell-O, a curve simply hangs. There is nothing so simple in the higher dimensions, but the idea that the singularities of a mapping indicate some irregularity or change in its behavior carries over intact. So does the play between differentiation and the existence of singularities. A mapping between manifolds of the same dimension has a singularity at a point x just in case its derivative is singular at x. (To the fastidious intellect this definition will prompt alarming images of a regress looping backward: f is singular at x if its derivative is, and its derivative is singular at x only if – need one be told? – the Jacobian of f at x is singular at x. But the last step ends the regress. The Jacobian matrix of a derivative is singular if its determinant is 0.) This yet leaves unexplained the connection between the Jacobian of f and f itself; the missing link is put in place by the Inverse Function Theorem, which describes the extraordinary relationship between the local diffeomorphic structure of a function and the properties of the determinant. Thus if the Jacobian of f is singular, the diffeomorphic mapping around x that f would otherwise effect collapses. In sending U onward, where U is a neighborhood of x, f must *concentrate* the mapping – onto a single point, perhaps. And here f encounters a singularity.

Simple though this system is, it makes for a powerful theorem in elementary analysis. If $N = \mathbb{R}$, the set of regular values of f is dense on the line: Its complement is of measure 0. This is Sard's theorem. In showing that the singular values of a differentiable function are crushed on the line, Sard's theorem offers some interesting evidence that the ideas thus far introduced, which may at first sight be mistaken for the trifling treasures of the elementary calculus, have a kind of brute and careless power; they suggest the sinister family of heavily oiled automatics, which go off with a roar and leave things generally in a shambles after they have been fired.

The conspiratorial pattern that stealthily links points and curves in $\mathbb{R}^2$ reappears after a fashion in $\mathbb{R}^n$ as well. It is the Hessian, as I have said, that goes in for the second derivative of a map in higher dimensions. Consider, in this regard, a function f. At x, I assume, the Jacobian determinant of f is 0. If the matrix that serves to express the Hessian of f at x is *invertible* (a mathematical operation akin to a complete tumble in gymnastics), x is *non-degenerate*; if not, x is *degenerate*. Degeneracy is obviously a cousin to the condition that the second derivative of a function be set at 0; but concavity and convexity are properties of curves and not maps; there is no simple-minded carry over of concepts from $\mathbb{R}^2$ to $\mathbb{R}^n$.

Mighty Morse

Morse theory is the creation of the American mathematician Marston Morse. By all accounts a rather shy and endearing figure – a picture published in a volume commemorating his birthday shows him looking up at the camera, pink and Presbyterian – Morse thought early in his life to analyze the calculus of variations in the large, and stayed doggedly with this topic ever after. In his obituary notice, Steven Smale talks disapprovingly of Morse's mathematical insularity, and, while this is not decisive evidence in his favor, it is very hard to ignore.

By a Morse function mathematicians now mean a smooth function $f: \mathbb{R}^n \rightarrow \mathbb{R}$, whose singular points are non-degenerate. $\mathbb{R}^n$ might well be replaced by a differential manifold, but the Morse functions, wherever they start, end on the line. This division of real-valued functions is based on the behavior of the Hessian and superficially suggests the case of curves in $\mathbb{R}^2$ classified by means of their second derivative. But so far, the classification is without content: It is not obvious that the Morse functions have any strong properties save for the fact that they are Morse functions.

But, of course, they do. The theorem that makes this plain comes in three parts. In the first place, Morse demonstrated that the Morse functions are locally stable at their non-degenerate singular points. Since, by definition, Morse functions have no degeneracies, such functions are stable at each of their singular points. So long as the Morse functions are not too sparse, this is a striking bit of good news. Happily, the Morse functions are dense in $C^\infty(\mathbb{R}^n, \mathbb{R})$ – the space of all functions from $\mathbb{R}^n$ to the real line, a certain natural topology prevailing. This marks the theorem's second step. And, finally, Morse showed that at their singular points the Morse functions could be expressed in a canonical or normal form. If x is a singular point of f, there exists a number λ such that in a neighborhood of x, and after a suitable change of coordinates, $f(x)$ is just

$$(f)x = x_1^2 + \ldots + x_\lambda^2 - x_{\lambda+1}^2 - \ldots - x_n^2$$

Such is Morse's theorem, which is unaccountably known in the literature as Morse's lemma. A global version of the lemma is also true, and carries just a bit more punch. For a family of functions f: $M \rightarrow \mathbb{R}$, where M as a smooth n-dimensional manifold replaces $\mathbb{R}^n$, only the Morse functions are globally stable, and then only when their singular points take distinct singular values. The set of such Morse functions is open and dense, and hence generic, in $C^\infty(M,\mathbb{R})$.

There is local Morse, then, and global Morse as well. Between them Morse theory comes woofing to all my themes. Stability figures here, and so does genericity; both stability and genericity turn on the singularity structure of certain maps. Any function suitably near a Morse function *is* a Morse function: The singularity structure of *all* Morse functions is completely determined by degeneracy conditions on *any* Morse function. The Morseness of Morse functions is contagious. Not only are functions close to Morse functions Morse functions in turn, but other functions can always be approximated at their singular points by Morse functions. If f is degenerate at x, some Morse function f^* stands close to f at x in the C^∞ topology. The singularity structure of all maps from $\mathbb{R}^n$

or M to $\mathbb{R}$ is thus frozen in place by the Morse functions, a vivid example of action at a distance.

Thom's Way Out

Differential topologists turned early in the history of their subject to the stable singularities; the degenerate singularities got short shrift simply because they were degenerate. But with the classification of the Morse functions complete, the mathematician is yet incapable of treating such simple functions as $f(x) = x^3$, which is petulantly unstable at $x = 0$, where 0 is a degenerate singularity of f. Functions near x^3 have very different qualitative properties.

The discussion – and the history of mathematics to this point – now comes to a temporary halt at the intersection of a dream and a dilemma. The dream is quite simply to extend the classification of the stable maps to cover the degenerate singularities; the dilemma is just that it cannot be done. Caught thus, the mathematician's immediate inclination is to wriggle upward toward higher ground – resolution by enlargement. In this regard, René Thom's striking and original idea was to deal with the degenerate singularities by embedding f in a *family* of functions – $f_a = x^3 - ax$, for example. Families of this sort he dubbed *unfoldings*, blossom-like constructions in the theory of maps. The unfoldings themselves are classified by a simple count of their parameters – 1 in the present case, r in general. (This number must not be confused with the *dimension* of the function, which is again 1 in this example but n in general.) In this family of functions, a new function arises from each change in the value of a. The focal point of the mathematician's attention has now been changed. The idea of stability carries over to the case of a family of functions, and so does the definition of what it means for families of functions to be equivalent.

Yet what of the catastrophes in all this? The general and intuitive core to the concept of a catastrophe is simply the idea of discontinuous change. I am speaking now of the public concept that Thom both appropriated and defined. The mathematical concept is parasitic. Like instantaneous velocity in physics, catastrophes lie on the porous border between mathematics and the other sciences. Birth and death are the great examples in biology; marriage and divorce, in social life. But catastrophic changes are not simply changes that are sharp. Catastrophes occur in stable ways. Each human being lives, and then he dies. The passage from life to death is stable in that everyone, so far as I know, goes over to the other side in much the same way, and, indeed, follows much the same sequence: The Big Chill followed by The Big Sleep. Only the details vary. The general case is one in which individual differences are wiped away in a blur of family resemblances.

In mathematics, Thom argued, the correlate to a catastrophe occurs when the number of minima (or maxima) in a family of functions changes abruptly in a stable fashion. There are in this idea precisely three mathematical concepts. Those discontinuities that mark catastrophes in biological or social life get expressed by the mathematical idea of a discontinuity; stability acquires its full (if ambiguous) meaning as mathematical stability; and the catastrophes are reflected from *within* mathematics by considerations that pertain to the behavior of a family of functions.

The theory of singularities is marked by a single great and controlling idea: Classification, if it is to be accomplished at all, must be achieved locally, at the singular points. It is there, and only there, that things change. Where there is no change, there need be no classification either. Let me call all functions that agree locally with a given function near a given point, the *germ* of the function. A catastrophe is made up of a pair consisting of the germ of a smooth real-valued function taken at a degenerate singular point, and its unfolding or immersion into a family of functions. To speak of the unfolding is to highlight the *family* of functions; to speak of an immersion, the *passage* from the unstable functions to their unfoldings.

The dialectical play between unstable functions and their unfoldings may be easily illustrated by considering the tension between $f(x) = x^3$ and $f_a = x^3 + ax$. This is an example that I have already mentioned. At 0, $f(x)$ is unstable. f_a, on the other hand, is a catastrophe-liable object: The number of minima in f_a changes abruptly as a itself crosses from 1 to -1. When a is greater than 0, f_a has but one minimum; and none at all when a is less than 0. But here is the extraordinary and subtle point: While the germ of f is *un*stable at the origin, f_a, when considered as a family of functions, is not, and this makes all the difference in the world.

Like that elephant being appraised by a congregation of the blind, a great mathematical theorem appears in the mathematical world in any number of effective guises. The more one feels, the more one sees. In what follows, I bring this little discussion to its dramatic close by redescriptively stressing that aspect of Thom's theorem that represents Thom's debt to Morse. The unstable germs, Thom demonstrated, form a *finite* family of just 11 types. Each unstable germ may be universally unfolded in certain characteristic ways. The theorem's live, twitching nerve, however, may best be exposed by considering retrospectively the classification of various stable maps. The points to some space are, I imagine, suspended like so many glowing lights. At a regular point, no analysis is required – mathematician, pass by. At a non-degenerate singular point, stability is much a matter of Morseness: Either the stable functions *are* Morse functions, or they may be expressed by Morse functions. Until now, the degenerate singular points have remained unresolved, with functions in their ken behaving drunkenly. There is nothing that can be done to redeem the unstable functions; they are what they are. Any one-parameter *family* of functions, however – and here my voice blends polyphonically with the voice of Thom's theorem – may be expressed by means of the stable family of functions $x^3 + ax$; any two-parameter family of functions, by a family of functions of the form $x^4 + ax^2 + bx$. This process goes upward, but only to a finite extent. At each level, the unstable germs pass placidly into the stable unfoldings.

This is the big picture; what is yet missing is the big punch, which now follows. The analysis takes place at a degenerate singular point. Loitering about that point are various families of functions. The reader is invited to pick one at random. That family, Thom demonstrated, is generically likely to be equivalent to an unfolding. Those unfoldings take a variety of characteristic forms – the *fold*, the *cusp*, the *swallowtail*, the *parabolic umbilic*, to employ the poetic and somewhat corny terminology that Thom introduced. Now, in every family of functions, *two* important numbers figure. The first records the dimension of the function; the second, the number of its parameters. Toward the dimensionality of the function, Thom's theorem is astonishingly indifferent: It holds for any value. But the theorem is valid only for those parametric values that are less than or equal to 5. In this sense, the theorem is at once expansive (it covers any value of n) and constrained (it applies only to a finite restriction of r). And, finally, the point behind the punch: The unfoldings form a generic and stable set of maps in the space of all smooth maps.

Stable Shapes

Plane mappings take the plane into the plane; so do pairs of equations, and any mapping f: $\mathbb{R}^2 \rightarrow \mathbb{R}^2$ may also be represented in Cartesian coordinates $x = u$ and $y = v$, where x and y section one plane into quadrants, and u and v section the other. The mapping $x = u$, $y = v$ leaves the plane unchanged; $u = x^2$, $v = y$ folds one half of the plane onto the other. This is the subject treated by the American mathematician Hassler Whitney in a majestic and important paper.

Plane mappings represent the transmutation of one flat surface to another, as when a waiter's towel is folded on itself, or a handkerchief sectioned into quadrants. Arising out of the plane, the mappings disappear back into the plane. Mathematics is always easier when there is something more substantial to see. The plane mappings are interesting in that their real effect appears somehow *between* a pair of planes, somewhere in mid-air. When the waiter folds his towel across his arm with a jaunty wave, what sticks in the mind is the motion of the mapping, and not the mapping itself. To a certain extent, and by an abuse of rigor, this effect lends itself to graphic expression, as in Figure 1, which captures a mapping on the wing.

In folding the plane over itself, the topologist creates a mapping that is intuitively stable. The plane may be rolled along its fold between the thumb and forefinger; under small perturbations, it stays the same in its overall shape. A knife passed cleanly through the folded plane will make a cut along a parabola whose tip just touches the axis. The parabola is a stable shape and has a singularity at the origin. The plane as a folded whole is but the sum of such parabolas, and a singularity extends along the entire fold.

How many such stable shapes are there when the plane is mapped onto the plane? Morse theory suggests

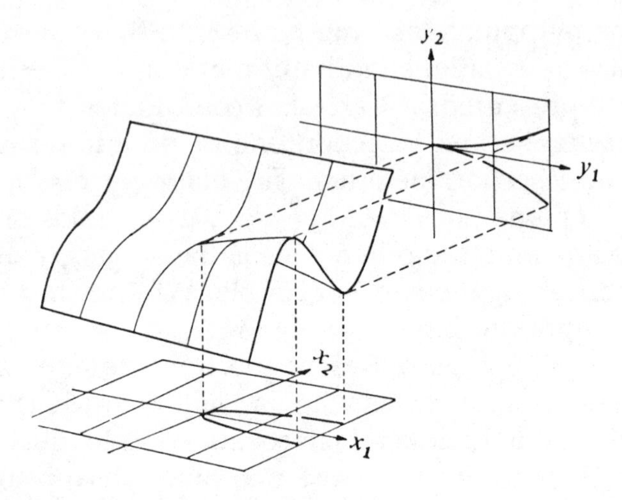

Figure 1.

that in this regard the singularities are crucial. At the singular points to a mapping information is concentrated and dense. In its insistence that the singularities of a function be analyzed by reference to the function's derivative, Morse theory maintains a fond and familiar family connection to the concepts of the calculus. To these concepts Whitney brought the full resources of mathematical generality. The derivatives of a mapping he endowed with the character of an abstract space – what is now called the *Jet* space. It points are made of jets – there is a certain undeniable logic to this terminology – and the jets themselves fixed by the Taylor expansion of mappings truncated at a certain point – the kth term in the expansion, say. The k-jet of f at $x + h$ is thus a polynomial in h; the Jet space is the set of such polynominals. The simple condition of non-degeneracy reappears now as a constraint on the Jet space. If the first partial derivatives of a mapping vanish, the second partial derivatives of the mapping may find themselves on one of two imagined sides to the Jet space. To the left, is the "bad" set; to the right, the "good" set. So long as the second partial derivatives fall to the right, one has non-degeneracy. There is nothing here that Morse might not have mentioned, but the overall effect is cleaner and more spacious.

Imagine now that f maps an open subset of the plane onto the plane. At a point p, f is either singular or regular – this as before. The point p is *good* just so long as the Jacobian of f at p, taken in at least one way, is not 0; f itself is good if every point in $\mathbb{R}^2$ is good.

The singular points of f, Whitney argued, form a family of smooth curves in the plane. This is obvious enough from the example of the fold, whose singular points lie touching the axis in a way that suggests a soft hammock. Such curves Whitney called *general folds*. The lemma that establishes their existence uses nothing more elaborate than the Implicit Function Theorem.

Thus defined the general folds seem pretty flabby. But behind the general folds lie the *normal* folds. These are Ur-folds, as it were, anthropological archetypes in the theory of plane mappings. I would much prefer to call the normal folds *canonical*, chiefly for the ominous double rumble of the word, but "canonical folds" evokes a whiff of the canonical forms of Morse theory and so signals a common mathematical urge.

Take f as before, and let p be any singular point of f – the spot where its behavior changes. If f is good its singular points form a smooth curve and a general fold C passes through p. Suppose that $\phi(t)$ is a smooth parameterization of C at p such that $\phi(t) = p$. Whitney stipulated that p is a *fold point* of f if

$$D_p f(\phi) = 0,$$

and a *cusp point* if

$$D_{pj}f(\phi) = 0; \qquad D^2_{pj}f(\phi) \neq 0.$$

This makes for a two-way classification of the singular points. Finally, a point p of f is *excellent* if it is a regular, fold, or cusp point; and f is itself excellent if every point in $\mathbb{R}^2$ is excellent.

In terms of concepts, there are but four so far. The division of mappings into those that are regular at p and those that are singular at p comes first. Then there is the play between good and bad mappings. The classification of singular points into folds and cusps is next. The concept of an excellent mapping is last, but the argument points to the set of such mappings like a pistol. These are the canonical mappings.

I spoke just now of an argument, but what has passed so far is more like a delicate tissue of definition. If the normal folds are to act as archetypes in the theory of plane mappings, they should have dramatically strong properties. Stability is important here, and so is genericity. Without them the definitional design created thus far remains incomplete.

Whitney worked things in every which way. The excellent mappings are both stable and dense. Stability is not itself a term that figures in Whitney's paper, but this is a matter merely of terminology. What Whitney *did* show was that the excellent mappings could be represented in terms of smooth coordinates (x,y) and (u,v). The definition of fold and cusp points owes nothing to the particular choice of parameters, so this comes to the same thing as local stability. To say that a smooth coordinate system may be introduced around p and $f(p)$ in terms of which the excellent mapping f takes a characteristic form is just to say that f is resistant to perturbation. The regular mappings leave the plane unchanged: $x = u$, $y = v$; folds take one flap of the plane onto the other: $u = x^2$, $v = y$; and cusps swirl the plane about a point: $u = xy - x^3$, $v = y$.

This is important, of course. It means that the excellent mappings are solid enough to count. But density counts for more if the image of an archetype is to survive. Here, too, Whitney was entirely successful. If f is any mapping of a subset of the plane onto the plane, then arbitrarily near f there is an excellent mapping f^*. So only the excellent mappings are locally stable, and arbitrarily near any mapping there is an excellent mapping. The fold and the cusp arise out of the plane as the only stable structures amidst an infinity of possible ghostly and impermanent deformations.

The Long Goodbye

That winter I lived in a little village to the south of Paris. Bures-sur-Yvette consisted of a pair of streets set at right angles to one another and arranged so that when one had walked past the two cafés, the grocery store, the *charcuterie*, and the *pâtisserie*, the streets themselves seemed to trail off irresolutely into the distance in a way that suggested that the village as a whole retained only enough energy to afford the illusion of reality over distances less than, say, half a mile.

I took my meals chiefly at the *Café des Sports*; breakfast was a matter of coffee and a croissant, of course; lunch was a confused Boschian babble in which one evil-tempered waitress would endeavor simultaneously to serve at least thirty workmen wedged uncomfortably on wooden benches, four men to a bench. Every now and again, I would take my evening meals at the town's only real restaurant, the *Crêperie*. It was always empty. The tile floor caused one's footsteps to reverberate unpleasantly. On the walls there was a series of pictures of sad-eyed waifs holding guitars out like crutches. There were candles on the tables. The restaurant was run by a very handsome, middle-aged woman and her stolid, unmoving, unexpressive son. I would enter the place in the early evening and seat myself. Within moments, that son would swim silently to my table. "*Vous avez choisi?*" he asked in the same toneless and monotonous voice each and every time. Writing with painful industriousness, he would inscribe my order (*crêpes* – what else?) in a little black notebook, and after gliding off gloomily, tear off the page and hand the sheet to his mother, who with studied deliberation affixed it to a post in the kitchen. Then he sat by the cash register, staring into space.

The *Institut des Hautes Études Scientifiques* lay across the railway tracks and up a little hill. To enter the institute, it was necessary to cross a park by means of a delicate doubtful ramble. The park was always empty and covered in winter with wet leaves. A whole family of charming little rodents (voles, I think) had taken over the premises as a private preserve and scampered indignantly through the leaves whenever I appeared. Once or twice I managed to walk silently enough to surprise a congregation at one or another of their meetings; with no place to run, and with *me* there looming over them like some nightmare (walking upright, large as the moon, and breathing fire from the hole in the center of its face), they would lie stock still, their bodies flattened to the ground, their sides trembling, tiny malevolent eyes gleaming.

I shared my offices at the Institute with a Polish mathematician who spoke no English and who regarded French with grim determination, his vowels, when he spoke, arranging themselves in a melodious rumble, his consonants acquiring, by means of a kind of linguistic transmigration, an extra syllable or a weird displacement of emphasis. His wife, curiously enough, had managed in the six months that they had lived in France to attain almost a pure, bird-song French; when she came by the office, *he* would remain still as the door knob, while she would chatter away, until he would silence her with a slavic glower and the demand, expressed I imagine in

Polish, that she speak in some language he could understand properly.

In the late afternoon, tea was served at the Institute. This constituted the day's only social event. A sullen, none-too-clean housekeeper would lay out the tea, the cream, and the coffee in beautiful silver pitchers. A platter of tasteless crackers of the sort that the French imagine are much favored by the English she then placed in the center of an intricate white doily. I generally came early to read the newspapers. By and by a few members of the Institute would appear: my Polish roommate, a few mathematicians from Germany, an American mathematician named Leo Rubble, whose only topic of conversation was his own research results. Sometime later, the great eminences of the Institute – René Thom, Pierre Deligne, David Ruelle, perhaps an important visitor from abroad – would make their appearance. I remember especially Pierre Deligne moving with great stateliness toward the tea. He was tall, with a curious, expressive face, a very high forehead, slanted guppy eyes, and a rapidly bobbing Adam's apple. After acquiring his tea cup, he would stand by the table and fix his gaze into space so as not to catch anyone's eye. The other eminences would then get their tea and crackers, one after the other, and then arrange themselves in the room so as to maximize the distance among them.

After tea I would walk back to the village. The light had already grown low and flat; those voles were safe in their burrows. At the *Café des Sports*, the waiters were busy stacking the chairs onto the tables for the night and bustling about industriously. The *Crêperie* was not yet open. At the *Librairie*, the gray-haired, faded, elegant woman who presided over the town's only bookstore sat at her counter, looking over the great treasures of French culture: Racine, Molière, Voltaire, Gide, Proust, and astonishingly enough, Raymond Chandler in a French translation – *Le Grand Adieu*.

Owing to Whitney

In thinking about differential topology, I find unaccountably that I favor anatomical (rather than historical) images. The relationship between Morse and Thom is one of frank nervous innervation, with the incompleteness of Morse's program acting intellectually as a kind of tense trigger; the relationship between Whitney and Thom is much more a matter of solid symmetry of structure, as when the anatomist displays to a roomful of gagging students the musculature of the cat in a way that suggests dramatically the points of contiguity between those sheathed and striped muscles and the muscles of the human shoulder.

Whitney's great theorem, I have stressed, is unusual in the degree to which it seems to erase the normally impregnable line between pure and applied mathematics. The stable shapes that arise when the plane is mapped to the plane are not purely mathematical objects; they have an existence in the real world. The retina of the human eye, for example, constitutes a plane of sorts, and what we see under ordinary circumstances is the flat and bounded surface of objects in space. Sight is therefore very much a matter of mapping between two planes, a fact that is of importance not only in the analysis of vision but in the interpretation of the graphic arts. In one of its many incarnations, Thom's classification theorem is best understood as a far-reaching generalization of Whitney's work. The stable shapes that are thus uncovered and made real and robust are neither purely mathematical nor entirely physical; like the fold and the cusp, they seem to occupy a region of the imagination that corresponds almost perfectly to those forms of which Plato spoke. In any event, the process by which mathematics is applied to the external world is always exciting, involving as it does the miraculous reinterpreation of the abstract in the concrete, as when a soul is said to leave its trace on the physical world in the form, say, of those dripping eyes on the Chicago Madonna, or when the causes of an effect, and the effect itself, are related by means of a function. This is the domain of elementary catastrophe theory.

There is animal aggression, for example, a subject made popular by Konrad Lorenz in a series of studies, and then adapted to mathematical purposes by Christopher Zeeman. What a dog does is pretty much a matter of what he wants or wishes to do; but suppose, just to grease the argument, that a dog's behavior may be resolved into two separate and conflicting tendencies. When faced with adversity, the beast may choose either to flee from his troubles or to fight them. Fleeing and fighting are represented by points in a space of *one* dimension. Those factors that get him to do what he does admit of an equally simple resolution. He is prompted, I am assuming, either by rage or fear. Rage and fear are represented by points in a space of *two* dimensions.

Behavior, I have said, is determined by two factors; I have not said how. On Zeeman's account, the relationship between cause and effect is mediated by a measure of *probability*, indicating, for various combinations of rage and fear, the *likelihood* that the behavior of the dog will be given by a particular choice of effects. I summarize the chief points. Rage in the absence of fear is likely to prompt an animal to fight; fear in the absence of rage is likely to prompt the same animal prudently to flee the scene. In the absence of either rage or fear, the animal's behavior is neutral: He simply stands there in dopey bewilderment. In the presence of both rage and fear, the animal's behavior is *ambiguous*: He may either flee or fight.

The mathematics of this matter, then, is just this: There are two spaces, C and X, say (representing, respectively, causes and effects), and one function f (representing probabilities). Variations in C give rise to *different* functions: f is parameterized by C and thus embodies the behavior of a *family* $F_{a,b}$ of functions.

As the example of animal aggression indicates, the systems studied by elementary catastrophe theory are both natural and very common. In two dimensions, they arise whenever a doubled pair of causes give rise to a single measured effect, with the relationship between cause and effect expressed by a function.

The mathematician now supplants entirely the psychologist, who, in any event, figures in this discussion only as a straight-faced straight man. Two causes of behavior have been introduced. The real world thus contracts to a two-parameter family of functions. Knowing that two parameters figure in the family, the mathematician yet knows next to nothing about the family itself; he is very much in the position of a man who remembers that he spent an unforgettable evening with two women without knowing what they looked like. Suffering thus from a form of intellectual amnesia, the mathematician recalls Thom's theorem. The standard unfolding of a two-parameter family of functions is $F_{a,b} = \frac{1}{4}x^4 + \frac{1}{2}ax^2 + bx$. (I have introduced numerical parameters to make the calculation easier.) In the jostling congregation of *all* functional families, the form of *this* family is typical. The mathematician interested only in the *local* circumstances influencing animal behavior might as well allow his own enigmatic family of functions to be replaced in all further calculations by its standard unfolding.

And now for another piece in the puzzle. The function f is of interest only in those cases in which f signifies a qualitative change in the probability of the animal's behavior. In Zeeman's model of animal aggression, f indicates the *most* likely course of action. Qualitative changes the mathematician identifies with the singularities of a function. To reflect *all* such singularities, the derivative of $F_{a,b}$ is set to 0: The result is the equation $x^3 + ax + b = 0$.

Now equations, I have stressed, give rise in analytic geometry to curves, surfaces, hyper-surfaces, and manifolds. $x^3 + ax + b = 0$ describes a cubic surface – M, to give it a name. A graph in two dimensions represents the values of the function by which it is generated; a surface does as much in three. M functions as a collocation of singular points, and singular points indicate a change in the behavior of a function. It is for this reason that M is dubbed a *castastrophe manifold*.

The analysis thus far has made use of two spaces, one function, and a surface. These items may be resolved into a single image, as in Figure 2.

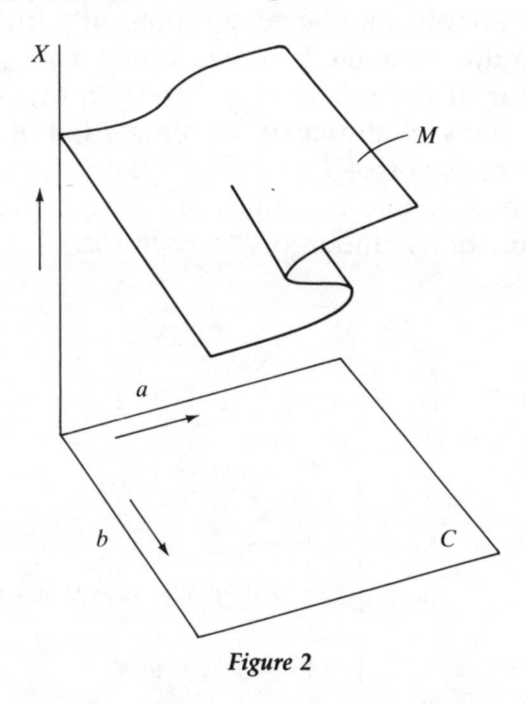

Figure 2

Imagine, now, that the castastrophe map has been projected downward so that each point on the castastrophe manifold is associated with a point on the plane below. With the lines of the projection inscribed, plain Figure 2 becomes potent Figure 3.

The peculiar cusp revealed on the plane will come as no surprise to the clue-minded reader. The catastrophe manifold contains a living record of all singular points in *f*. Its pleat, however, which undulates softly over itself, records along its lines of curvature those singular points in *f* that are degenerate. Degeneracies indicate the local emergence of an unfolding. But *f*, understood now as a mapping, traces a complete path from one plane to another. Whitney's theorem establishes that in such mappings, the fold and the cusp are paramount.

With this clue in place, it becomes plain that the cusp contains the crucial information about $F_{a,b}$ itself. Those points that are outside the cusp represent particular families within the complete family of functions $F_{a,b}$ in which the behavior of the family stays pretty much the same. When the lines of the cusp are crossed, however, the qualitative behavior of $F_{a,b}$ changes. Functions that previously had only one maxima now have two. Here the relevant functions undergo a *catastrophe*. A sharp and

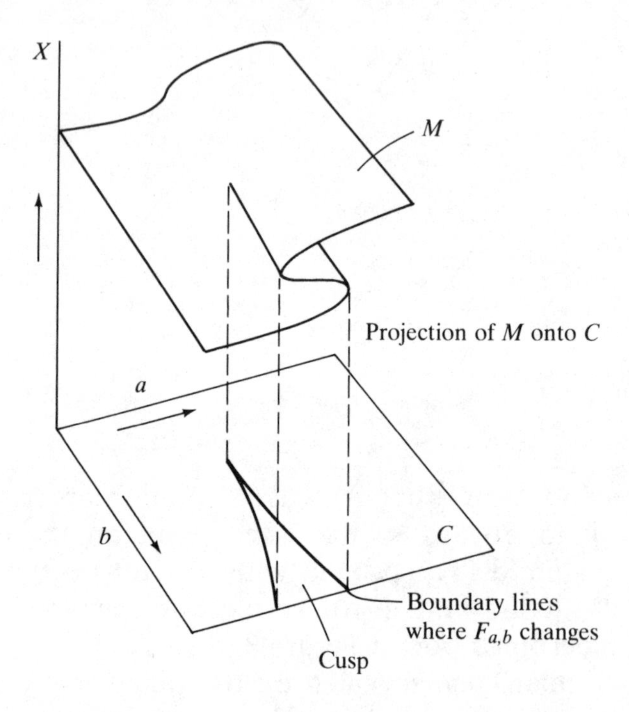

Figure 3

unassailable change in behavior is recorded. The entirely trivial implications that this fact has for animal behavior, the reader may figure out for himself.

The fold and the cusp are the prerogatives of Whitney's theorem, which coincides meticulously and miraculously with Thom's classification theorem in two dimensions. For the rest of the story, the mathematician trudges upwards. Instead of the two-dimensional example just scouted, I turn a tired hand to the *general* case of a family f of functions – one in which f varies both in the number of its parameters and in its dimensionality. The concepts that I have already introduced, I now transfer to this discussion. Thus suppose that M_f is a catastrophe manifold (or surface). Let X_f be the downward projection of this manifold. And let F be the set of *all* functions that like f end on the real line. Thom's theorem admits of expression in four parts. Assuming that the number of parameters (r) in f is less than or equal to 5 (a great, tragic, unavoidable assumption), there is an open and dense – and hence generic – set of functions F^* in F, with the following properties:

1) The catastrophe surfaces to which they gave rise are manifolds: When viewed locally, they look Euclidean;

2) The projection map of any function f in F^* is locally stable with respect to small perturbations; and

3) – the theorem's main mighty point – *Any* singularity in the projection map X_f is equivalent to one of a *finite* number of types – the *elementary* catastrophes.

The number of such singularities depends on r (and not n!) in the following way:

Table 1

r	1	2	3	4	5	6	7
elementary catastrophes	1	2	5	7	11	∞	∞

This restatement of Thom's theorem adds no new information, of course; those stable and generic families to which I have just alluded are nothing but the unfoldings that I have just discussed.

•••

This is the end to a little line of thought. The sympathetic reader who has followed the discussion to this point has noticed, no doubt, that while the central idea of Newtonian mechanics may be explained glibly, the machinery involved in the exposition of catastrophe theory is rather more burdensome. And, indeed, it is. Yet this way of looking at things, while not entirely wrong, is not entirely right either. Modern science commences, I think, at precisely the moment in which a physical concept receives a mathematical voice. In the case of Newtonian mechanics, that physical concept is instantaneous velocity; its mathematical voice is heard when the concept of the derivative of a function is made manifest. There is a good deal more to mechanics, of course; the elegant analytical mechanics of, say, D'Alembert or Hamilton goes quite beyond the simpleminded exposition that I have offered. And, yet, there it is: When everything is brushed away, and the thing is reduced to its absolute essentials, it is the derivative that lies at the heart of the mechanical world view. Once it has been introduced all else follows.

Within Newtonian mechanics, the parabola, the ellipse, and the hyperbola function as archetypes: They are the curves by which continuous change is classified. In catastrophe theory, there are the elementary catastrophes; they serve to impose a classification on discontinuous change (with what success I have not said). And here, too, there is a secret heart of sorts. The concept that informs the development of catastrophe theory is that of the singularity of a smooth map, and singularities are themselves mapped and measured by reference to the behavior of the derivative of a function. So in the end, the progression from simpleminded physics to the very sophisticated mathematics of Thom's theorem represents stages in the enlargement of a single idea, a kind of fantastic episode of barely controlled growth in the history of thought.

Classification and Its Discontents

The modern molecular biologist looks on life with a blunt dissective passion. This is his strength. It is also his weakness. Biological systems, Thom correctly insists, are not merely biochemistry made palpable. The Yak exists as a geometrical figure. Its life can be traced as the figurative sum of a series of cross-sections, where the cross-sections themselves are understood as images of the animal frozen in a particular geometrical configuration. Such cross-sections, tied together to form a temporal arc, exhibit striking discontinuities. Where the Yak ends there is Yaklessness, in the sense that the Yak, like any other creature, has a spatial boundary within which he is uniquely himself, yakley, as it were, and in the larger sense that the Yak comes into existence at a given time and goes out of existence at another. Even these time-bound catastrophes, which from the Yak's-eye point of view are so disagreeable, are themselves part of a geometrical schema, an extended pattern of movement in space and time, something like a majestic periodic wave in any number of dimensions in which the perishable individual appears briefly as one fragment of an endless figure.

Thom's grand objective is a theory of such forms, with the generation and change of biological structure mirrored in theory as aspects of geometry. Biological change is in its most important aspect sharply discontinuous. At the margins of Yak, whether spatial or temporal, the Yak simply *stops*. This suggests the pull of two contrary tugs, with stability covering those aspects of life that stay the same, and change, introduced now as a short term standing for concepts that need to be made precise, covering those aspects of life that do not. The coarse features of experience thus require for their analysis mathematical concepts that stay stable as they switch states. Those unfoldings at the center of the classification theorem come to mind here with a whoosh; they figure in the technical development of Thom's theory of models.

For all of its originality, the root to Thom's theory – the central ganglion from which all nerves radiate – may be traced backward toward a point of controversy in the development of complex function theory, the most beautiful (and sterile) of all mathematical disciplines. These nineteenth-century disputes and disputations are not my concern; I mention them only to elict one voice – that of Bernhard Riemann, who argued (against Cauchy and Weierstrass) that what counts in complex function theory are the singularities. In this sense, Thom is adapting to philosophy an idea of long standing in mathematical analysis.

In quite another sense, however, Thom's program makes contact with a certain somewhat shifty tradition in physics. Physicists classify as *phenomenological* theories at the surface of things (the theory of Fermi surfaces, for example, or the Landau theory of turbulence). These theories tend to provoke the physicist's displeasure or despair. They are insufficiently deep and suggest, when they work, only the occurrence of a lucky accident. Yet there is in the history of Newtonian mechanics precedent for a frankly phenomenological approach to the dynamics of moving bodies. The differential topologist sent into space (by whim or witlessness) along a hyperbolic orbit, and the luminous planets above him in the night sky, are treated alike as point masses in mechanics; their internal structure collapses to a fictive point and then vanishes.

A classification achieves a certain measure of conceptual economy only by associating objects that coincide in just some of their properties. Objects that agree in all respects are self-identical. Newtonian theory accomplishes its spectacular effects by classifying all objects in motion in terms of just three curves (the parabola, the hyperbola, and the ellipse). The classification of moving bodies is indifferent to their differences. But the example provided by this scheme is obviously of little help to the theoretical biologist who means to account for the laws of morphogenesis. In that context, *which* properties should count toward a classification?

The inductive intellect looks on things with a theoretical vengeance. All swans are white, it observes; it therefore makes no sense to account for their whiteness on a swan-by-swan basis. With grapefruits and basketballs, on the other hand, it is the other way around. Grapefruits are round because that is the way of all fruit; basketballs are spheres because nothing else in three dimensions bounces. There is no explaining their shape in anything like a common way. Yet in catastrophe theory, objects *are* associated on the basis of their form. Noticing that a grapefruit and a basketball describe similar shapes in space – they are both spheres – the catastrophe theorist, in classifying them together, risks coming to the conclusion that they are both nourishing and good to eat. The mathematician is, of course, not committed to a classification based on the tritest of geometrical properties. The objects that *he* assimilates will be surfaces or manifolds. But the sceptic may still find room to scruple.

The most ordinary groupings in nature arise in the most natural way. Things are classified by virtue of their sex or race or color or mass or valence. Such systems are successful largely because they are in such perfect accord with the theories that explain them. Thom's system of classification, by way of contrast, is alien. Objects in the biological world are classified by means of their geometrical properties, and then only with respect to their singularities. The mathematician, of course, finds this a source of satisfaction. "What is interesting about morphogenesis," Yung-Chen Lu remarks, "locally, is the transition as the parameter varies, from a stable state of the vector field X to an unstable state and back to a stable state by means of the process which we use to model the system's local morphogenesis." But as the chess master might observe, a scheme of this sort tends to identify a great many processes that otherwise seem distinct. What is interesting about chess, I hear him murmur wickedly, is the opening gambit and the final move that ends in mate. These correspond to the singularities of the system – the transition from an initial stable state (all pieces symmetrically arrayed) to an unstable intermediate state (play in progress) and back to a state made stable by the fact that one player can no longer move. *Au Roi*!

In writing about butterflies, Vladimir Nabokov remarked that the delight is in the details. In one respect, science is passionately particular. The physicist, on the other hand, persists in the view that great secrets are revealed only when things and processes are grouped together. In another respect, science is indifferent to diversity. The sceptic not much interested in physics may wonder why the ordinary world and its delights should be dismissed so enthusiastically. The answer, I suppose, is just that physics, like anything else, involves a trade-off, with the world and its wonders given up for something else – an overview or underview, insight, intellectual precision, comprehensiveness. Intellectual life is forever a matter of those who wish to worship at the veil of appearances and those who would brush it aside. The physicist generally insists that the world be carved and classified by means of the distinction between its macro- and microscopic states. The mathematician points instead to the distinction between a system's local and global states and strategies. However much he may argue with the physicists about the right classification to be imposed on the natural world, Thom is with them in one respect: When it comes to the veil of appearances, he is prepared to give that curtain a good solid yank. On Thom's system of classification, for example, *all* two-dimensional discontinuities find themselves as members of a common class. Their form Thom explains by means of the cusp; the cusp, by means of their form. The obvious differences between discontinuities in nature, Thom is prepared to ignore. But there is a penalty associated to this position. Those processes that the mathematician assimilates on the basis of their form must also be *explained* on the basis of their form. They have nothing else in common. If the mathematician responds with a good-natured shrug that items so collected should receive different explanations, as in the case of the basketball and the grapefruit, the sceptic is entitled to ask why they are collected together at all. The theoretical biologist, I suspect, may, like the chess master, come to insist that each game or gonad is exactly what it *is* and nothing else. The mathematician, adrift thus on geometry, needs geometry simply to stay afloat.

Centro di Physica

Travel in Italy in the company of a ravishing, almond-eyed, caramel-skinned young woman, given to wearing tight sweaters and slitted skirts and smiling up at waiters, taxi-drivers, and strangers in the street, and you will discover the perfect irrelevance of every human quality save beauty.

Paris had been elegant, gray and gloomy. We had looked with admiration at Sainte Chapelle, half in shade, half in light as the result of restoration on the stained glass, and Notre Dame, and then walked along the Seine, where barges were moving slowly downstream, and scuffed the chestnut leaves that were gathered in piles; but entering Italy from France was like steping from a cool autumnal patio into a radiant, multi-colored, high-summer piazza.

I had been invited to spend two weeks at the *Centro di Physica* in Trieste. In the mornings, I would deliver my lectures in a magnificently equipped auditorium, a microphone draped over my shoulders, an array of imposing buttons at my fingertips. These were supposed to raise and lower the blackboard. My audience consisted of exceptionally bright students of mathematics or physics from various third-world countries. Unfortunately, their English was all of the My-uncle-has-received-a-wire-from-Bombay variety; the more that I attempted to simplify my syntax, the more complicated it became, until in the end I found myself virtually submerged beneath a string of subordinate clauses, my students looking up at me with ill-concealed malevolence. Afterwards, I would repair to the leafy cypress-bordered terrace, and stand in the warm Adriatic sunlight, and wait and smoke.

By and by a taxicab would arrive. The rear door would open at a tantalizingly acute angle; I could see Victoria leaning over the front seat, pressing an uncounted wad of lire into the driver's moist hands. Then she would swivel and emerge, dressed in that throat-catching sweater she wore all through Italy, a narrow grey skirt, sheer hose, and fire-engine-red shoes. Poised between her seat and the square, she would hesitate for just a second and flash to the open air a smile of warm, glowing sunniness, while the taxi driver, who had been taking all this in from his rearview mirror, would lick his lips, and smack the steering wheel with his palm, and roll his eyes upward, as if to say that *Mamma Mia* this was all too much to bear.

In the afternoon, we would amble along the narrow beach, or look at the stone-heavy castle by the institute, or wander through a very elegant, empty park with wooden benches, tall drooping trees, and a view of the sea.

In the evening, we went out for dinner – Victoria and I and the mathematicians John Casti, Christopher Zeeman, and Charles Muses. From our table on the terrace of the restaurant, I could see the Adriatic, smooth in the gentle glow of early evening, and, on the far shore, the faint lights of Yugoslavia. The chef, a stout, friendly young man, with lustrous eyes and oiled hair, served us himself – course after course of regional specialities: eels in a wonderful white sauce; crustaceans and scallops; a dish of fragrant black mushrooms, served with an acrid white wine available only on the northern Adriatic coast; a sole in aspic, crusty bread, heavy in the center; artichokes, their hearts pureed; pasta; grilled meats; salad; and a dessert of which I remember only its golden, glowing color.

Twilight turned into evening. On the Adriatic, the fishing boats bobbed, their lanterns spreading a soft fuzzy light. An insect commenced a steady businesslike chirp from beneath the terrace. The air was balmy in the gentle dark.

The cheese came and went, and, finally, an Amaretto. Christopher Zeeman withdrew a mouth organ from his pocket and began to play. We had eaten and drunk a good deal.

"It's all a question of timing," Charles Muses said, answering a question that no one had asked.

I thought of time rolling on through things like a great wave, cresting here, dipping there.

Some cognac? No? Perhaps one of these biscuity cookies? asked the chef.

As if it were descending a series of steps toward darkness, the evening grew closer. A strange exotic night-bird began a deep, throat-clearing *haruumph* in the distance.

The waves of time were now washing up in Yugoslavia.

"I'll drink to that," said John Casti, who by now was prepared to drink to anything at all.

PART V

Doctor No

James Darnell, the chairman of the department of biological sciences at Columbia University, was a man in his mid-fifties, with silver hair and the high chubby cheeks of a squid. He sat at the very edge of his chair, elegant and erect, and held my résumé fastidiously with his fingertips, as if it carried a sample of sputum. It was a bright, clear, cold, cutting day late in the fall. On Broadway and Amsterdam Avenues, small clouds of grit and dust collected in mysterious eddies in the harsh sunlight, spun in the air, and then dissolved. Darnell's office was on the seventh floor of Schermerhorn Hall, at the end of a long unlit corridor. Like most of the busy men on campus, he had no use for undergraduates and chose his office to discourage their visits. The windows had never been opened, let alone washed, and the sunlight cut through the dirt and dust on the windowpane to fall into the room in long, silky, silent shafts.

I had been bounced as a management consultant from Flyte and Company earlier that year, and bounced again as a budget analyst from the city's smarmy Human Resources Administration just six months later. My enthusiasm for welfare work having been broken by a procession of waddling welfare mothers, I decided, for no good reason that I can now recall, to learn something about molecular biology. Perhaps it was the lovely, liquid swoop of the word "molecular" that prompted me. My training had been in analytic philosophy, a subject not known for the light it sheds on other subjects, and, indeed, like a black hole in space, not generally known for shedding any light whatsoever. But the great unversities of America were yet the wonderfully dreamy, open, and generous institutions they had always been. Campus homosexuals, for example, had recently declared themselves outraged by the university's failure to provide them a suitable lounge. With the froth of consolation on his lips, the President welcomed a delegation of deviates to his door: Horsefaced and solemn, he assured them of his every consideration. I figured I could pull down an assistantship in biology, my training in philosophy notwithstanding.

Darnell finally finished his scrupulous reading of my résumé and looked up at me. His eyes were pale, almost milky.

"I suppose," he said in a low cultivated voice, the consonants blurry, "that there really is a limit to how stupid a man with a Ph.D. can be."

I nodded and then stood and edged backward toward the door. I later learned that Darnell had given up private practice as a physician in order to devote himself to research. His speciality was carcinoma. Even sitting, he carried himself with a consecrated air.

A System of Belief

The thesis that theoretical biology constitutes a kind of intellectual Uganda owes much to the theory that biology is itself a *derivative* science, an analogue to automative engineering or dairy management, and, in any case, devoid of those special principles that lend to the physical or chemical sciences their striking mahogany luster. Naive physicists – the only kind – are all too happy to hear that among the sciences physics occupies a position of prominence denied, say, to horticulture or agronomy. The result is *reductionism from the top down,* a crude but still violently vigorous flower in the philosophy of science. The physicist or philosopher, with his eye fixed on the primacy of physics, thus needs to sense in the other sciences – sociology, neurophysiology, macramé, whatever – intimations of physics, however faint. This is easy enough in the case of biochemistry. Chemistry is physics once removed; biochemistry, physics at a double distance. Doing biochemistry, the theoretician is applying merely the principles of chemistry to living systems. His is a reflected radiance.

In 1831, the German chemist Friedrich Wohler synthesized urea, purely an organic compound – the chief ingredient in urine, actually – from a handful of chemicals that he took from his stock and a revolting mixture of dried horse blood. It was thus that organic chemistry was created, an inauspicious beginning, but important, nonetheless, if only because so many European chemists were convinced that the attempt to synthesize an organic compound would end inevitably in failure. The daring idea that *all* of life – I am quoting from James Watson's textbook, *The Molecular Biology of the Gene* – will ultimately be understood in terms of the "coordinative interaction of large and small molecules" is now a commonplace among molecular biologists, a fixed point in the wandering system of their theories and beliefs. The contrary thesis that living creatures go quite beyond the reach of chemistry biochemists regard with the alarmed contempt that they reserve for ideas they are prepared to dismiss but not discuss. Francis Crick, for example, devotes fully a third of his little monograph, *Of Molecules and Men*, to a denunciation of vitalism almost ecclesiastical in its forthrightness and utter lack of detail. Like other men, molecular biologists evidently derive some satisfaction from imagining that the orthodoxy they espouse is ceaselessly under attack.

Curiously enough, while molecular genetics provides an interpretation for certain Darwinian concepts – those differences between organisms that Darwin observed but could not explain – the Darwinian theory itself resists reformulation in terms either of chemistry or physics. This is a point apt to engender controversy. Analytic philosophers cast reduction as a logical relationship. Given two theories, the first may be reduced to the second when the first may be *derived* from the second. The standard and, indeed, the sole example of reduction successfully achieved involves the derivation of thermodynamics from statistical mechanics. In recent years, philosophers have come to regard the concept of direct reduction with some unhappiness. There are problems in the interpretation of historical terms – the Newtonian concept of mass, for example – and theories that once seemed cut from the same cloth now appear alarmingly incommensurable.

To speak of the *logical* structure of biological thought is at least tentatively to suggest that biological theories have something even vaguely discernible *as* a logical structure. This represents, I think, the lingering influence on the philosophy of biology of standards current in the philosophy of physics. Now theories in physics quite often are logically disorganized and, despite the animadversions of philosophers, none the worse for that. Their intellectual robustness is, I think, a function of the fact that physical objects are almost always entirely determined by physical theories. To the philosopher who wishes to know what a quark *is*, the physicist need only point to the laws that describe quarks – the principles of quantum chromodynamics, say. The philosopher who insists that this is all very well, and demands to know further what a quark *really* is, has asked an unanswerable question. In the case of life, however, the objects scouted by biological theory have an antecedent conceptual existence, one that is quite indifferent to expression in theoretical terms. These objects are fixed in our imagination by their position within a dense matrix of concepts, with the matrix itself animated by dim, inarticulate biological throbbings. In comparison to physical theories, biological theories are circumscribed; the philosopher asking innocently for an account of life is hardly in a position to dismiss on principled grounds any number of possible answers – a play of biochemical forces, physics in its most complex state, the coordinative interaction of large and small molecules (Watson's answer), aspects of the Mind of God, the structures forged to protect the gene, the appearance in the universe of pity and terror. It is some measure of the confusion in contemporary thought that each of these answers seems roughly right.

But no matter the degree to which molecular biology is logically disorganized, the definition of reduction that I have cited is incomplete as well as irrelevant. In Mendelian genetics, the concept of a gene is theoretical, and genes figure in that theory as abstract entities. To what should they be pegged in molecular genetics in order to reduce the first theory to the second? DNA, quite plainly, but how much of the stuff counts as a gene? "Just enough to act as a unit of function," argues Michael Ruse, a philosopher whose commitment to prevailing orthodoxy is a model of steadfastness. But in biochemistry the notion of a unit of function is otiose, unneeded elsewhere. To the extent that biochemistry is molecular genetics, it does not reflect completely Mendelian genetics; to the extent that it does, it is not biochemistry, but biochemistry beefed up by a conceptual padded shoulder.

What holds in a limited way for molecular genetics holds in a much larger way for molecular biology. Concepts such as *code* and *codon*, *information*, *complexity*, *replication*, *self-organization*, *regulation*, and *control* – the items required to make molecular biology *work* – are scarcely biochemical. The biochemist following some placid metabolic pathway need never appeal to them.

Population genetics, to pursue the argument outward toward increasing generality, is a refined and abstract version of Darwin's theory of natural selection, one that is applied directly to an imaginary population of genes. Selection pressures act on the molecules themselves, a high wind that cuts through the flesh of life to reach its buzzing core. Has one here achieved anything like a reduction of Darwinian thought to theories that are essentially biochemical, or even vaguely physical? Hardly. The usual Darwinian concepts of fitness and selection remain unvaryingly in place. These are ideas, it goes without saying, that do not figure in standard accounts of biochemistry, which very sensibly treat of valences and bonding angles, enzymes, fats, and polymers – anything but fitness and natural selection.

To the standard conditions on reduction, then, I would add a caveat: No reduction by means of inflation. The Darwinian theory of evolution is the great, global, organizing principle of biology, however much molecular biologists may occupy themselves locally in determining nucleotide sequences, synthesizing enzymes, or cloning frogs. Those biologists who look forward to the withering away of biology in favor of biochemistry and then physics are inevitably neo-Darwinians, and the fact that this theory – *their* theory – is impervious to reduction they count as an innocent inconsistency.

Theoretical biologists still cast their limpid and untroubled gaze over a world organized in its largest aspects by Darwinian concepts, and so do high-school instructors in biology – hardly a group one would think much inclined to the idea of the survival of the fittest. Unlike the theory of relativity, which Einstein introduced to a baffled and uncomprehending world in 1905, the Darwinian theory of evolution has never quite achieved canonical status in contemporary thought, however much its influence may have been felt in economics, sociology, or political science. If mathematical physics offers a vision of reality at its most comprehensive, the Darwinian theory of evolution, like psychoanalysis, Marxism, or the Catholic faith, constitutes, instead, a system of belief. Like Hell itself, which is said to be protected by walls that are seven miles thick, each such system looks especially sturdy from the inside. Standing at dead center, most people have considerable difficulty in imagining that an outside exists at all.

Scopes Two

In 1982, for example, biblical literalists in Little Rock, (cousins kidnapped at birth by Gypsies to those fierce-eyed Methodists who in 1923 went down for the full count against Clarence Darrow), forced the State Supreme Court of Arkansas to appraise the constitutionality of a new law requiring the Darwinian theory and biblical accounts of creation to be exhibited together in the state's public schools, like competing midget tag-team wrestlers. The American Civil Liberties Union, ever outraged, filed a brief with the court contesting the law. Photographs of Judge William Overton showed him leaning over the bench, heavy brows furrowed in concentration, a grim look in his eyes suggesting that whatever the law might say, *he*, at any rate, was not going to get caught on the side at which history snickers. The ACLU brought in the heaviest guns that it could find: Harvard's Stephen Jay Gould, bearded and puffy, dressed in three-piece suits, like an investment banker; the University of California's Francisco Ayala, sleek as a ferret and almost as fast on his feet; Yale's Harold Morowitz, fuzzy-faced, dark, saturnine, scowling.

For the most part, the witnesses ranged against the Darwinian theory were a scruffy lot. Nothing in the scientific circles in which they had moved – biblical schools, institutes for creationism, such stuff – had prepared them for the vigor of an academic attack. By and large, the Darwinians romped home. They stayed clear of Chandra Wickramasinghe, though, who came to Little Rock, apparently, to argue that both sides to the controversy were sunk deeply in error. An astrophysicist by training, and hence superior in status to any of the biologists, Wickramasinghe compared the Darwinian theory in point of plausibility to the thesis that a Boeing 747 might arise spontaneously from a junkyard struck by a tornado. This is not the most elegant way of putting things, but it has a certain primitive force. The account of creation offered in the Bible, Wickramasinghe dismissed with a snort before his supporters reminded him just who had paid for his round-trip ticket from God-knows-where to Little Rock. Astrophysicists generally know a good deal of fancy mathematics of precisely the sort that biologists scrupulously avoid, so The Wick spent most of his time in Little Rock on a leash, a deep-voiced bull mastiff without anything to bite. In the end, the Darwinians agreed that the Darwinian theory was quite beyond controversy among reasonable men; but why then were they there, explaining once more to an audience of bewildered and resisting laymen how it is that time and chance work their will in an ancient world?

Employed Again

The golden glitter of the day had gone completely by the time I got to see Barry Honig. A large clot of undergraduates, all aching anxiety over glands or grades, stood outside his door. The office looked over the Stalinist squares of the university's new law complex. From the window I could see that long shadows already stood on the beige, pre-stressed concrete pedestrian walkway on Amsterdam Avenue. There were several squat, gray-steel desks in the room, and an odd collection of secretarial chairs of the sort that tilt backward and have a tendency to tip over. Honig was sitting with his feet on the desk in front of him, eating pistachios and spitting the shells with no great success toward a green metal trash can. He looked to be in his late twenties or early thirties. Slouched as he was, his thick shoulders gave him a look of imperturbability. When he stood, the great, filled, half-globe of his belly flopped down into the pouch formed by his pale-blue button-down shirt. "How you doin'?" he asked, and motioned me to a chair with a fluid sweep of his outstretched hand, the palm upward, the fingers stiffened. His thin lips, very dark and almost florid coloring, and bushy eyebrows gave to his face a curiously Oriental cast. His cheeks still held the red rubble of adolescent acne.

The office lights had been left off; we sat and talked comfortably as the afternoon slipped finally into twilight and the traffic hooted and whistled in the streets below.

James Darnell had sent my résumé ahead of me. A copy, I could see, lay on the desk, coffee-stained and smudged. Like almost all academics, Honig had no taste for formal interviews. "I suppose I'd better ask you something," he said glumly, but then he couldn't think of anything to say. He did deliver himself of the opinion, held tenaciously by most scientists, that analytic philosophy is largely nonsense – "What do you guys do all day long? Diddle with each other?" – but in my heart of hearts I agreed with him then, and I am not sure that I don't agree with him now.

Haunting History

Charles Darwin completed his masterpiece, *On the Origin of Species*, in 1859. He was then forty-nine, ten years younger than the century, and not a man inclined to hasty publication. In the early 1830s, he had journeyed to the islands off the Peruvian coast aboard H. M. S. *Beagle*. The stunning diversity of plant and animal life that he saw there impressed him deeply. Prevailing biological thought had held that each species is somehow fixed and unalterable. Looking backward in time along a line of dogs, it is dogs all the way. Five years in the Pacific suggested otherwise to Darwin. The great shambling tortoises of the Galapagos, surely the saddest of all sea-going creatures, and the countless subspecies of the common finch, seemed to exhibit a pattern in which the spokes of geographic variation all radiated inward to a common point of origin. The detailed sketches that Darwin made of the Galapagos finch, which he later published in *On the Origin of Species,* show what caught his eye. Separated by only a few hundred miles of choppy ocean, each subspecies of the finch belongs plainly to a common species, and yet, Darwin noted, one group of birds had developed a short, stubby beak, another, living northward, a long, pointed, rather Austrian sort of nose. The variations among the finch were hardly arbitrary: Birds that needed long noses got them. By 1837 Darwin realized that what held for the finch might hold for the rest of life, and this, in turn, suggested the dramatic hypothesis that, far from being fixed and frozen, the species that now swarm over the surface of the earth *evolved* from species that had come before in a continuous, phylogenetic saxophone-slide.

What Darwin lacked in 1837 was a theory to account for speciation. The great ideas of fitness and natural selection evidently came to him before 1842, for by 1843 he had prepared a version of his vision and committed it to print in the event of his death. He then sat on his results in an immensely slow, self-satisfied, thoroughly constipated way until news reached him that A. R. Wallace was about to make known *his* theory of evolution. Wallace, so far as I know, had never traveled to the Galapagos, sensibly choosing instead to collect data in the East Indies; considering the same problem that had earlier vexed Darwin, he had hit on precisely Darwin's explanation. The idea that Wallace might hog the glory was too much for the melancholic Darwin: He lumbered into print just months ahead of his rival.

The theory that Darwin proposed to account for biological change is a conceptual mechanism of only three parts. It involves, in the first instance, the observation that small but significant variations occur naturally among members of a common species. Every dog, for example, is doggish in his own way. Some are fast, others slow, some charming, others suitable only for crime. Yet each dog is essentially dog-like to the bone, a dog, *malheureusement,* and not some other creature. Darwin wrote before the mechanism of genetic transmission was understood, but he inclined to the view that variations in the plant and animal kingdoms arise by *chance,* and are then passed downward from fathers to sons.

The biological world, Darwin observed, striking now for the second point to his three-part explanation, is arranged so that what is needed for survival is generally in short supply: food, water, space, tenure. Competition thus ensues, with every living creature scrambling to get his share of things and keep it. The struggle for life favors those organisms whose variations give them a competitive edge. Such is the notion of *fitness.* Speed makes for fitness among the rabbits, even as a feathery layer of oiled down makes the Siberian swan a fitter fowl. At any time, those creatures fitter than others will be more likely to survive and reproduce. The winnowing in life effected by competition Darwin termed *natural selection.*

Working backward, Darwin argued that present forms of life, various and wonderful as they are, arose from common ancestors; working forward, that biological change, the transformation of one species to another, is the result of small increments that accumulate across the generations. The Darwinian mechanism is both random and determinate. Variations occur without plan or purpose – the luck of the draw – but Nature, like the House, is aggressive, organized to cash in on the odds.

Cashing In

By and by we ran out of things to say. Whatever his views on analytic philosophy, Barry Honig was a man of instinctive and even spontaneous generosity; before the silences between us deepened uncomfortably, he had offered me a position as his research assistant. I had no set responsibilities under the scheme of things that he envisioned, which was fine with me, but I would have to learn some organic chemistry, and then some biochemistry. It might be a good idea, Honig added, if I picked up something of quantum mechanics, but this he could teach me himself. My salary would come from the Ford Foundation. How so? I associated the Ford Foundation chiefly with the unlucky McGeorge Bundy and any number of dim charities of the sort in which bright pairs of American volunteers distribute colorful condoms to the fecund and uncomprehending natives of some dusky village. I certainly had no objections to standing in line before a trough of such grandeur. It just seemed a little odd.

A long story, Honig said, his reedy baritone deepening to a conspiratorial bass. The Ford Foundation had recently experienced a sharp twitch along the corporate nerve controlling environmental concerns. Money had been spent on air pollution and the tender and easily bruised oceans, the speciality just then of the lisping Jacques Cousteau, and more money spent studying the upper atmosphere, where chemists, working backward, of course, had traced a sinister connection between the depletion of the ozone layer and the use of deodorant sprays, but somehow the effects on the environment of new chemical compounds had gone unstudied. Rachel Carson had written brilliantly of chemical pesticides in *Silent Spring*, but no one, it appeared, had since turned an attentive hand to the problem. Cyrus Levinthal was one of the senior members of the department of biological sciences. He had come to Columbia University from the Massachusetts Institute of Technology in a swap for six sociologists. A physicist by training, Levinthal had done important work at MIT on the measurement of DNA; now he was studying the nervous system of something small-scaled and squishy – a worm, I believe. Honig had conceived an ox-like admiration for Levinthal: "The man is always thinking," he remarked in a way that suggested that Levinthal's brain was like Jerry Lee Lewis in a state of permanent and ineradicable hyperexcitability. Whatever his intellectual virtues, Levinthal was obviously one of those luminous creatures perfectly designed by some obscure Darwinian mechanism to excel at grantsmanship. The Ford Foundation having developed an interest in chemical pollution, however vagrant, Levinthal hit on the idea of creating an international registry of all chemical compounds manufactured in the United States, Europe, and the Soviet Union. The registry would be computerized, of course, since fifty thousand new chemical compounds are produced each year; computers, too, would analyze the molecular structure of each compound with an eye toward determining their potential for harm. No sooner had officials at the Ford Foundation breathed in than Levinthal told them what they were about to say; by the time that they were able to breathe out, he had gotten away with the loot. Most of the money was earmarked for expensive hardware – computer graphics and curious gadgets of every sort – but there was enough money left over to support a handful of research assistants.

I understood, I said, but then it was really time to go. The seventh floor of Schermerhorn Hall was empty when I left Honig's office – except for the mute and monstrous biological specimens that stood collected in their glass cubicles. The ancient elevator, which clanked and wheezed coming up from the basement, looked when it stopped as if its heavy and graffiti-covered doubled doors were a mysterious nictitating membrane opening to a final sightless eye.

Misprints in the Book of Life

With the exception of Shirley Maclaine, apparently, everything that lives, lives but once. To pass from fathers to sons is to pass from a copy to a copy. This is not quite immortality, but it counts for something, as every parent knows. The higher organisms reproduce themselves sexually, of course, and every copy is copied from a double template. Bacteria manage the matter alone, and so do the cells within a complex organism, which often continue to grow and reproduce after their host has perished, unaware for a brief time of the gloomy catastrophe taking place around them. It is possible, I suppose, that each bacterial cell contains a tiny copy of itself, with the copy carrying yet another copy. Biologists of the early eighteenth century, irritated and baffled by the mystery of it all, thought of reproduction in these terms. Peering into crude, brass-rimmed microscopes, they persuaded themselves that on the thin stained glass they actually saw an homunculus. The more diligent of the biologists then proceeded to sketch what they seemed to see. The theory that emerged had the great virtue of being intellectually repugnant. Much more likely, at least on the grounds of reasonableness and common sense, is the idea that the bacterial cell contains what Erwin Schrödinger called a *code script* – a sort of cellular secretary organizing and recording the gross and microscopic features of the cell. Such a code script would logically be bound to double duty. As the cell divides in two, it, too, would have to divide without remainder, doubling itself to accommodate two bacterial cells where formerly there was only one. Divided, and thus doubled without loss, the code script would require powers sufficient to organize anew the whole of each bacterial cell. The code script that Schrödinger anticipated in his moving and remarkable book, *What is Life?*, turns out to be DNA, a long and sinewy molecule shaped rather like a double-stranded spiral. The strands themselves are made of stiff sugars, and stuck in the sugars, like beads in a sticky string, are certain chemical bases: Adenine, Cytosine, Guanine, and Thymine, A, C, G, and T, in the now universal abbreviation of biochemists. It is the alteration of these bases along the backbone of DNA that allows the molecule to store information.

One bacterial cell splits in two. Each is a copy of the first. All that physically passes from cell to cell is a strand of DNA. The message that each generation sends faithfully into the future is impalpable, abstract amost, a kind of hidden hum against the coarse wet plops of reproduction, gestation, and birth itself. James Watson and Francis Crick provided the correct description of the chemical structure of DNA in 1952. They knew, as everyone did, that somehow the bacterial cell, in replicating itself, sends messages to each of its immediate descendants. They did not know how. But the chemical structure of DNA, once elaborated, suggests irresistibly a mechanism for both self-replication and the transmission of information. In the cell itself, strands of DNA are woven around each other, and by an ingenious twist of biochemistry matched antagonistically: A with T, and C with G. At reproduction, the cell splits the double strand of DNA. Each half floats for a time, a gently waving genetic filament; chemical bonds are then repaired as the bases fastens to a new antagonist, one simply picked from the ambient broth of the cell and clung to, as in a singles' bar. The process complete, there are now two strands of double-stranded DNA where before there was only one.

What this account does not provide is a description of the machinery by which the new cells are actually organized. To the biochemist, the bacterial cell appears as a small sac enclosing an actively throbbing biochemical machine. What the machine extrudes are long and complex molecules constructed from a stock of twenty amino acids. These are the *proteins*. The order and composition of the amino acids along a given chain determine which protein is which. The bacterial cell contains a complete record of the right proteins, as well as the instructions required to assemble them directly. The sense of genetic identity that marks *E. Coli* as *E. Coli*, and not some other bug, is thus *expressed* in the amino acids by means of information *stored* in the nucleotides.

The four nucleotides, we now know, are grouped together in a triplet code of sixty-four codons, or operating units. A particular codon is composed of three nucleotides. The amino acids are matched to the codons: C-G-A, for example, to arginine. In the translation of genetic information from DNA to the proteins, the linear ordering of the codons themselves induces a corresponding linear ordering first onto an intermediary, messenger RNA, and then onto the amino acids themselves – this via yet another messenger, transfer RNA. The sequential arrangement of the amino fixes finally the chemical configuration of the cell.

Molecular biologists often allude to the steps so described as the *central dogma*, a queer choice of words for a science.

The dour Austrian monk Gregor Mendel founded the science of genetics on purely a theoretical notion of the gene. In DNA one looks on genetics bare: The ultimate unit of genetic information is the nucleotide. All that makes for difference, and hence for drama, in the natural world, and that is not the product of culture, art, artifice, accident, or hard work, all this, which is brilliantly expressed in perishable flesh, is a matter of an ordering of four biochemical letters along two ropy strands of a single complex molecule.

The central dogma describes genetic replication, but the concepts that it scouts illuminate Darwinian theory from within. Whether as the result of radiation or chemical accident, letters in the genetic code may be scrambled, with one letter shifted for another; entire codons may be replaced, deleted, or altered. These are *mutations* in the genetic message. They are arbitrary, because they are unpredictable, and yet enduring, because they are genetic. The theory by which Darwin proposed to account for the origin of species and the nature of biological diversity now admits of expression in a single English sentence: Evolution, or biological change, so the revised, the *neo*-Darwinian theory runs, is the result of natural selection working on random mutations.

Life Itself

In the first thirty years of this century, the great physicists – Einstein, of course, Dirac, Schrödinger, Heisenberg, the taciturn Bohr – moved from triumph to triumph with the calm inevitability and special sense of grace afforded sleepwalkers in the dead of night. Of Einstein, especially, it was said that physics simply melted in his mouth. Others may see things differently, but to my mind the various branches of physics, whether quantum mechanics or relativistic astrophysics, appear to be converging toward a babbling form of mysticism in comparison with which the theory of the orgone box is a very marvel of clarity and sound good sense. Einstein himself remarked late in life that the God of Physics – a figure, I imagine, much like Wotan – is subtle but not malicious. Einstein was vouchsafed an overwhelming vision of order early on; in the end, he acknowledged that what he had seen in a flash in the first flush of his young manhood was only a part of the truth, and he died in darkness, his work incomplete. This may not be evidence of malice in the scheme of things, but neither does it suggest much by way of divine helpfulness.

Thousands upon thousands of creatures swarm over the surface of the earth. Some are strikingly successful in the struggle for life. There are well over fifty-thousand varieties of insects; like podiatrists, most of them are thriving. Some species have engaged in veritable prodigies of adaptation, with the hominids moving from an opposable thumb to a disposable tampon in what amounts to the blink of a biological eye. Others, like the shark, never adapt to anything new at all and represent only a persistent cough in the long dark night of evolution. Living creatures grow, and they carry on ceaselessly, striving, acquiring, laying up provisions and storing them, cunning and resourceful; they reach out, impelled by some dim ancestral power, some muted voice heard with a musky throb of recognition by everything that lives, one to another; they mate and multiply, and then, unreconciled and, for all I know, unreconcilable, they sink into death, decay and oblivion.

More than any particular thing that lives, life itself suggests a kind of intelligence evident nowhere else; reflective biologists have always known that in the end they would have to account for its fantastic and controlled complexity, its brilliant inventiveness and diversity, its sheer *difference* from anything else in this or any other world.

An Experiment in Education

Honig was as good as his word. One morning he made an attempt to teach me quantum mechanics. Standing against the blackboard, his hands clasped together, he endeavored to summarize everything he knew of the subject in a single, sinewy sentence. With a pause for breath, he stopped, struck by the importance of a thought that seemed defiantly to resist inclusion in his scheme of things.

"You understand all about the exponential function, don't you?" he asked hopefully.

I said that I did. With a shrug that spoke of the

complexity of the matter, Honig then appealed to me with his eyes, as if he expected me – of all people! – to seize on that hidden connection between what he wished to say and what he had said, and which was obviously tormenting him like the tickle before a sneeze.

"You see?" he asked.

The Team

Organic chemistry I found a difficult, tormenting, itching, hair shirt of a subject. God, however, tempers the wind to the shorn lamb: Before too much time had gone by, Honig asked me to drop my studies and devote myself entirely to the international registry of chemical compounds. Within a week I discovered what was, in any case, obvious: No chemical company in the United States had the faintest intention of sharing proprietary information with members of the department of biological sciences at Columbia University; more often than not, when I wrote to Du Pont or to Monsanto or to Dow Chemical, I would receive my original letter back by return mail, an indignant "No!" slashed in red letters across the top. Representatives of the Soviet Union, to whom Levinthal presented his plan at the United Nations, treated his proposal as if it carried an embarrassing disease.

Much later in the fall, Levinthal called the entire team to his office. It was a mournful New York Saturday morning in a minor key. Steady rain had streaked the windows with long, soot-darkened spikes. Honig was there, and so was a fish-eyed computer technician whom Honig had hired. It was the first time that I had actually seen Levinthal at close range. He wore a white cardigan sweater and tucked his heels underneath him when he sat. His face was smooth and unlined, thin-skinned and delicate. We all reported our progress with deeply insincere enthusiasm. Fish-Eyes remained standing for the entire hour. The team may have gotten together early in winter, but I was never invited to another of their meetings.

Plenitude

Systems of belief, like kings and their kingdom, exist for a time and then become a part of the perpetual inventory of remembered things. No one knows how life on this planet originated. The prebiotic seas, so one story goes, resembled a thick and viscous soup, something, perhaps, like borscht. Great jagged flashes of lightning flickered in the night skies. Various molecules shifted beneath the surface of the oceans, and, quite by chance, some of them came together and stuck, forming, in time, ever larger primitive molecules. One thing led to another. In 1952, the American chemist Stanley Miller attempted to verify this account experimentally. He filled a test tube with various inorganic chemicals, shook the mixture up, and then passed sparks through the test tube in order to simulate the effects of lightning. When he opened the test tube, after allowing the thing to cook quietly in the sun for a week or so, he found some foul-smelling *organic* chemicals stuck to its bottom.

Still, it is difficult to know what to make of all this. Francis Crick – he of the Animadversions Against the Vitalists – has entertained the speculation that life on earth originated elsewhere in the universe and was simply sent – an organic telegram containing what is at best a mixed message. No sooner did Crick commit this masterstroke to paper, than Wickramasinghe, popping up in Cambridge, and Fred Hoyle, the luckless astronomer who staked his all on the steady-state theory of creation while everyone else backed the Big Bang, went one step further, arguing not only that life arose in space, but that it represents a form of cosmic intelligence.

Some creatures have six legs, others have twelve, and still others sixty-four. Certain fish have eyes that are like pinholes; others, eyes that stare simultaneously in two directions and are a marvel of bizarre complexity. Strange bladder-like swimming shapes fill the ocean waters, things flap in the night – an endless show, a stunning panorama, especially on television, with even the trees shifting their great load of leaves to face the light, everything different, it would seem, yet divided into companionable kingdoms, phyla, genera, and species, and built from the bottom up according to a common plan, evidence, surely, that on the molecular level, the Creator was interested in architectural economies.

Ancient and medieval philosophers, struck by the fabulousness of it all, never for a moment imagined that everything they saw stemmed from a single cell – some supremely lucky accident in the turbid prebiotic oceans. Behind the Visible World lies brooding the Invisible God; the various biological species He has sculpted separately from so much clay, breathing moistly to infuse them simultaneously with life. The species are not only separated in the obvious sense that a human being is remarkably unlike a toad (smaller lips, longer feet) – they occupy regions in the realm of biological possibility that are mutually inaccessible. There is in the nature of things no path between one species and another, no slow or sudden process made manifest in biology that would explain how creatures of one sort pass irrefrangibly into creatures of another sort – this because on the ancient and medieval view there is no growth or evolution within the biological kingdom, except for individual growth, and no change either. Everything that is actual is possible, and everything that is possible is actual, made into flesh, and divided into kinds. Such is the principle of *plenitude*, an idea that in ancient and medieval thought is inextricably linked to the cognate notion of continuity. The ideas of plenitude and continuity may not directly imply each other, but they are mutually reinforcing, and together they make for the conception of the biological universe as a part of the Great Chain of Being.

Static, immutable, unutterably pregnant, complete, continuous, unchanging and divine, the Great Chain of Being dominated the imagination of Western thinkers for well over one thousand years. Not a trace of this majestic idea remains in either popular or academic thought, and precisely the same evidence that suggested to the medieval schoolmen that the various species of animal and plant life on earth were fixed and permanent suggests to the modern scholar that they are in constant and ceaseless flux, growth, and evolution.

Good as Gould

Among other things, medieval thinkers believed that human beings were unique in ways that were absolute and inviolable. This doctrine many modern biologists have emphatically rejected. "The western world," Stephen Jay Gould remarks, "has yet to make its peace with Darwin." The great impediment to this reconciliation, he goes on to add in his mad way (the sense strong that he is urging a difficult truth on a dogmatic public),

> lies in our willingness to accept continuity with ourselves and nature, our ardent search for a criterion to assert our own uniqueness. Chimps and gorillas have long been the battleground of our search of uniqueness, for if we could establish an unambiguous distinction – of kind, rather than degree – between ourselves and our closest relatives, we might gain the justification long sought for our cosmic arrogance. Educated people now accept the evolutionary continuity between human and apes. But we are so tied to our philosophical and religious heritage that we still seek a criterion for a strict division between our abilities and those of a chimpanzee.

Now I quote all this not merely because Gould holds a chair at Harvard and I do not, although this made the target all the more tempting, but because Gould represents a charming intelligence corrupted by a shallow system of belief.

No distinction in kind rather than degree between ourselves and the chimps? No distinction? Seriously, folks? Here is a simple operational test: The chimpanzees invariably are the ones *behind* the bars of their cages. There they sit, solemnly munching bananas, searching for lice, aimlessly loping around, baring their gums, waiting for the experiments to begin. No distinction? Chimpanzees cannot read or write; they do not paint, or compose music, or do mathematics; they form no real communities, only loose-knit wandering tribes; they do not dine and cannot cook; there is no record anywhere of their achievements; beyond the superficial, they show little curiosity; they are born, they live, they suffer, and they die.

No distinction? No species in the animal world organizes itself in the complex, dense, difficult fashion that is typical of human societies. There is no such thing as animal *culture*; animals do not compromise and cannot count; there is not a trace in the animal world of virtually any of the powerful and poorly understood powers and properties of the human mind; in all of history, no animal has stood staring at the night sky in baffled and respectful amazement. The chimpanzees are static creatures, solemnly poking for grubs with their sticks, inspecting one another for fleas. No doubt, they are peaceable enough if fed, and looking into their warm brown eyes one can see the signs of a universal biological shriek (a nice maneuver that involves hearing what one sees), but what of it?

One may insist, of course, that all this represents a difference merely of degree. Very well. Only a difference of degree separates man from the Canadian goose. Individuals of both species are capable of entering the air unaided and landing some distance from where they started.

Ovid in his Exile

Schermerhorn Hall at Columbia University was the scene of many strange experiments. One day, a very young chimpanzee escaped from the building and, flushed with its freedom, began to gambol and frolic on the pathetic square of shabby and well-worn grass that served as a lawn in front of the building. A crowd quickly collected. The mathematician Lipman Bers joined me. A scruffy puppy noticed the commotion and scooted into the square where the chimpanzee was playing. The two animals promptly became friends, but the puppy, it soon became apparent, was less intelligent than the chimpanzee. Again and again he would find himself maneuvered into absurd and humiliating positions. "So stupid," snorted Bers, referring to the dog. Pleased and flattered by the attention, the chimpanzee began to refine his act and play to the crowd, using gestures, and even facial expressions – the universal rictus of triumph, for example – that everyone recognized. After a while, the chimpanzee's frantic owner, a rather dishy young woman, I recall, collared him in the courtyard and the game was over. As the chimpanzee was led away, he waved to the crowd, a true sportsman. The puppy sat on its haunches and panted assiduously.

I learned later from Bers that research biologists were trying to teach the chimpanzee the American Sign Language. They had been working with an older animal, but evidently the beast, while learning some signs, grew unsurprisingly to detest his owners, who finally shipped him to a zoo in San Diego. There he occupied himself unprofitably in attempting to teach the other animals to sign, a splendid case of the incompetent endeavoring to instruct the indifferent.

"A vast tragedy," Bers remarked sentimentally, "like Ovid in his exile."

I mention this sad little story only to remark on its ironic conclusion. For a time during the 1970s, a number of biologists were actually convinced that they had taught chimpanzees and great apes to talk; many of them reported long conversations, chiefly about bananas (*Me: More!*), that they held with their charges. Their research was no sooner published than it was accepted and believed, largely, I think, because a crude Darwinian theory – there is no other – made it difficult to imagine that profound and ineradicable differences exist between human beings and the rest of the animal world. Penny Peterson at Stanford, Herbert Terrace at MIT, and David Premack at the University of Pennsylvania all convinced themselves that somehow the great apes had sat in stony silence throughout the vast reaches of biological time only because they lacked human conversational companionship.

The inevitable, sceptical reaction soon set in. Videotapes taken of chimpanzees revealed, when carefully analyzed, that what had passed for chimpanzee conversation was nothing more than prompted signings in the best of cases – a record of the beast's pathetic endeavor to say whatever it was that his trainer wished him to say; in the worst of cases, the beast simply babbled (*More Me More More!*), his signs utterly devoid of meaning. Herbert Terrace, who had wasted years in browbeating the poor creatures, examined videotapes of his own encounters with his animals and came away shaken. Some work, of course, continues, but to little effect. Ever credulous, scientists now report that they have engaged the dolphin in stimulating conversation. Next year, no doubt, it will be the turkey.

Seventeenth-century Jesuits wondered why dogs do

not talk. Their conclusion bears repeating. They have nothing to say.

George

George Pieczenik mumbled a good deal when he talked, the words coming to the surface with the sound of cabbage being sliced. He was very thin, with a long, narrow, irregular face. No one in the department of biology quite knew what he did, but Barry Honig told me that Pieczenik was interested in the theory of evolution, and I found out from friends that he had a wonderful reputation for intellectual eccentricity.

We met one day in the library and then drifted off to Broadway for coffee. The campus was crowded with the usual run of drifters, undergraduates, professors in dingy, crepe-soled shoes, drug dealers from Harlem, and coeds from Barnard and City College, shuttling across campus, books pressed against the dense, untouchable mass of their wedged bosoms.

We crossed 112th street and finally settled into one of those coffee shops that are damp and smell of wet shoe leather in late fall. I mentioned the points about the theory of evolution that I found perplexing.

"It's a mess," Pieczenik said amiably.

We met regularly after that, in the library, in coffee shops, and in Riverside Park, which was then turning bare and bleak as the first winds of winter blew across the sullen waters of the Hudson River.

They were curious, those conversations of ours. From my point of view, molecular biology was a great graveyards of facts; Pieczenik saw the subject as a vibrant field of force, something almost alive, and sensuous in its capacity to respond to intellectual probing.

"Wha'dya think them microtubules do?" he would ask me.

I had no idea.

"See," said Pieczenik, "I have a *theory* about them microtubules," and, indeed, Pieczenik had a theory about virtually everything in biology and biochemistry. But the theories that he had, he circumscribed by means of his rock-solid laboratory technique. *I* would talk to him of mathematical models in biology, topological spaces, group theory, whatever came into my head. He would listen, staring dreamily off into space, two trails of smoke streaming from his nostrils, his lean, sharply angled face tilted upward. Thinking in that position, he would occasionally rub his lips with the tips of three fingers. "What does it mean?" he would finally ask, meaning, in turn, what does it mean *experimentally*.

I thought for a time that I might provide a mathematical framework in which our discussions might be embedded. I was prepared to do anything but work. I got as far as suggesting a natural measure of distance for the space of all polypeptides. I could never come up with anything that followed from this idea.

Pieczenik's own interests in biology nonetheless coincided neatly with my own, a circumstance that at the time I regarded as a point in his favor. He was convinced that the nucleotides respected certain *grammatical* constraints and naturally concluded that only a grammatical mechanism could account for this fact. I was interested in the structure of the grammar, and looked forward to a wonderful partnership in which he would do the messy and disagreeable work while I would claim the credit.

Pieczenik had actually made a significant discovery along these lines. Certain nucleotide triplets function in any biological system as genetic controls. These are the so-called *terminator* codons. When transcription has been completed, the terminators shut down the machinery of the cell, thus providing a hypothetical period to the genetic message. Suppose, however, that the mechanism slips by a single frame so that entirely the wrong triplets are being read or run. In examining specific sequences of DNA (which he had laboriously prepared), Pieczenik discovered that the genetic code contains a subtle mechanism by which frame misreadings of this sort may be prevented. As a result of misalignment, the molecular biological machinery of the cell comes to an automatic rumbling halt.

This is an interesting observation. The requirement that sequences of nucleotides be self-halting if incorrectly read dramatically shrinks the space of all possible proteins. In this regard, internal terminators function as universal principles of biological grammar. Still, as Pieczenik remarked, their existence was pretty much what one might have expected. "It's not *big*, Dave," he said. Far more striking was his discovery of palindromic sequences. A palindrome is any sequence of letters that mirrors itself along a point of symmetry: A B C D C B A, for example. In examining those same stretches of DNA (dog-eared, by now, to the point of disreputability), Pieczenik noticed that certain extremely long portions of the sequence were organized as palindromes, and exhibited, when examined in just the right way – not *my* way, needless to say – an astonishing internal symmetry.

In linguistics, a grammar comprises a finite set of rules generating all and only the sentences of a given language; in mathematics, an automaton is construed as an abstract machine. What struck me most about Pieczenik's palindromes was a point of classification. Like wrestlers, automata are ranked by their intellectual weight. At the top of the heap or hill, there are the Turing Machines, which are capable of doing *anything*, and thus occupy a position of lonely and brooding inscrutability; at the bottom, the finite-state automata, devices that are generally incapable of doing more than running elevators. Somewhere in between are the *push-down storage* automata. These devices correspond to phrase-structure grammars, and, indeed, machines and grammars represent two ways of expressing one concept. The push-down storage automata are distinguished from finite-state automata by a nice point: Only the push-down storage automata are capable of generating palindromic sequences. If molecular biological grammars exist, they must thus be at least as complex as push-down storage automata.

We talked on for a few more weeks that winter; later in the year, Pieczenik accepted an invitation to continue his research at the Molecular Research Council in Cambridge.

I received a few crabbed letters from him that spring. He was much occupied, he wrote, in straightening out Francis Crick on a few fundamental matters in molecular biology. After that I lost touch with him entirely.

Several years later, I spotted his picture in *Time* magazine. He had made an important discovery. One of his many theories had paid off. He was now a great man.

The Evidence for Evolution

Evolution, it is held, takes place over a very long time. The human hand evolved from the inhuman paw, step by step, one incremental improvement following another, a tortuous process, endlessly delayed, endlessly extended. No one, of course, has actually seen the whole business at work. The speckled butterfly, to take a famous example, was observed by British biologists in the north of England to change its wing coloring in order to maintain its mimetic protection. Laboratory insects, most notably the fruit fly *Drosophilia*, have been tracked through a series of evolutionary changes in wing structure and color. These anecdotal examples provide no real evidence that Darwinian mechanisms are at work and no evidence, surely, that Darwinian theories *explain* the process by which a new species arises from one that is old.

On this matter, the science of paleontology has some

bearing. Long periods of geological upheaval during the earth's early history have had the effect of trapping a variety of flora and fauna beneath the shifting tides of geological debris and freezing them there in an easily visible pattern. The record of fossils so laid down make up a *paleolithic stratum;* the various strata are read from the bottom up, and by moving upward, paleontologists can form a picture of the progression of organisms over time.

One might expect that the record in the rocks and the neo-Darwinian theory would fit, hand to glove, with the bottom layer exhibiting the simplest microorganisms, and each succeeding layer trailing off continuously into the next. In fact, the fit between theory and data is poor. The fossil record appears to end abruptly, well before the likely data at which life on earth commenced. Virtually no multi-cellular fossils have been found in the Precambrian rock strata; yet at the beginning of the Cambrian stratum, one sees an explosive proliferation of life forms – *de novo,* as it were, appearing abruptly, without obvious ancestors. There are gaps in the fossil record, strange discontinuities, an almost complete absence of intermediate forms, as if, to everyone's astonishment, the individual species had been sculpted separately, just as the medievals believed all along. The paleontological record is, of course, not evidence for special creation, but neither does it make for a great happiness among biologists. "Despite the bright promise that paleontology provides as a means for 'seeing' evolution," David Kitts has written in the journal *Evolution,* "it has presented some nasty difficulties for evolutionists, the most notorious of which is the presence of gaps in the fossil record. Evolution requires intermediate forms between species, and paleontology does not provide them."

Of course, if the fossil record does not fit the theory, it is always possible to adjust the theory to fit the record. In science, an enterprising theoretician has several degrees of freedom within which to maneuver before the referee reaches ten and the final bell comes to clang. Steven Jay Gould, who was trained as a paleontologist, surveyed the fossil evidence early in the 1970s and came to the obvious conclusion that either the theory or the evidence must go. What went, on his scheme of things, was the neo-Darwinian orthodoxy by which species change into different species by means of an endless series of infinitesimal changes, continuously, like the flow of syrup. Instead, Gould argued, biological change must have been *dis*continuous, with vast changes taking place at once. Such was his model of *punctuated equilibria*. It fits the fossil record far better (if it makes sense, even, to talk of scientific fit here), but the model achieves faithfulness to the facts only by chucking out the chief concepts of the Darwinian theory itself, and while paleontologists have been glad to have had Gould's company, evolutionary theorists have looked over what he has written with the cool, slitted, appraising glance of a butcher eyeing a sheep.

•••

In an entirely different way, some philosophers have always found something fishy in the Darwinian theory of evolution. An obvious sticking point is the concept of fitness itself. If by the fitter organisms, biologists mean merely those that survive, then the doctrine that natural selection winnows out those organisms that are not fit expresses a triviality. This is a logical point and not a matter of experiment or research. The biologist who wishes to know why a species that represents nothing more than a persistent snore throughout the long night of evolution should suddenly or slowly develop a novel characteristic will learn from the neo-Darwinian theory only that those characteristics that survive, survive in virtue of their relative fitness. Those characteristics that are relatively fit, on the other hand, are relatively fit in virtue of the fact that they have survived. This is not an intellectual circle calculated to inspire confidence.

Biologists, it goes without saying, reject with florid indignation the idea that the Darwinian theory is empty. Ernst Mayr, for example, asserts that "to say that this is the essence of Darwin's reasoning is nonsense." These are forthright words. Yet in writing of traits that have *no* apparent selective advantage, Mayr argues that "the mere fact that such traits have become established makes it highly probable that they are the result of selection." If this does not mean simply that those traits that survived survived, whatever else it might mean is very obscure indeed.

The evaporation of content from the notion of fitness is nowhere more apparent than in mathematical genetics, a subject with a reputation for pointlessness inferior only to that of modern tax law. In A. Jacquard's *The Genetic Structure of Populations,* the crude idea of fitness has disappeared in favor of the notion of "selective value," which Jacquard defines as a number "proportional to the probability of survival from conception to adulthood of individuals with [a given] phenotype." Whatever it is that accounts for the probability of survival, however, has become impalpable, a power registered only through its effects. This is rather as if an engineer were to define *stress* as a number proportional to the probability of failure, without ever inquiring why a particular beam buckled, or why beams buckle in general, or whether stress and structural failure represent two concepts or only one.

The doctrine that survival favors the survivors is what logicians call a *tautology,* a statement that is all form and no content. For obvious reasons, evolutionary biologists are uncomfortable with the idea that the chief claim of their theory is roughly on the intellectual order of the declaration that whatever will be, will be. Every organism, to continue the argument, needs to accomplish certain biological tasks simply in order to stay alive. The antelope must get food in order to keep on getting food, and it must avoid becoming food for those animals such as the lion that nourish themselves by eating other animals. In looking for food, there is an obvious advantage to claws, teeth, speed, keen sight, or a first-name familiarity with the headwaiter. Is this not, then, a step in the right – the Darwinian – direction? I am dubious. The obvious cases do appear to sort themselves out according to Darwinian principles. But were the pig to be born with wheels mounted on ball bearings instead of trotters, would it be better off on some scale of porcine fitness? No one knows, although some guesses are possible. The general idea that a biological organism requires certain powers if it is to survive is inadequate to explain the *specific* form those powers happen to take.

The petals of the orchid *Ophrys apifera*, for example, exhibit a design similar to the design inscribed on the genitalia of the female bee. The mimetic effect is quite striking, at least to the bee. But why such a complex mechanism, such fantastic drollery, when no other orchid requires anything like it?

The Pacific salmon, to take another example – the last – is a magnificent fish – long, heavy, silvery, sleek. It is spawned, among other places, in the streams that ultimately feed into Puget Sound and then wash out to sea. After a number of years spent in the open ocean, the salmon returns to its spawning beds to lay eggs and die.

In the late fall, as the thin sun casts wicked-looking cloud-stained streaks over the northern lowlands, thousands upon thousands of salmon begin to course through the inland sea, seeking out the mouth of the particular river that will ultimately wind backward through the evergreen forests to the very stream in which they were spawned. The geographic precision and unerring instinct that enables these fish to find their spawning grounds is largely a mystery. Where there is a means, there is a mechanism, and, presumably, some set of features enables the salmon to head home again – distinctive smells, perhaps, or subtle magnetic signals, or the tang and sheen of the air.

The journey undertaken by the Pacific salmon is unbearably poignant because it is at once desperate and directed toward death. This is to describe things in the abstract. To stand alongside the Duwamish or Steilacoom rivers in mid October, and watch the headwaters course with salmon, is to be struck by the unbearable grandeur of life itself. For the salmon, going backward along a river means going upward: over waterfalls, past rock formations, swimming all the time against the current and toward an ever-narrowing wedge of water. By the time the fish have pushed several miles upstream, they are exhausted. Their skins are grey, and mottled with parasite fungi. Their eyes are no longer clear. They expend energy in spurts, panting heavily when they come to still waters. Yet they are still frenzied, maddened by some vibrant call. Along the banks of the river, eagles sit patiently on the drooping branches, hardly bothering to fish. What they need will wash to shore. In the end, only a handful of fish make it to the headwaters of the rivers, and there they lay their eggs. But a handful is all that is needed, for each female deposits thousands of eggs, and of these thousands, thousands yet again will live to complete the cycle, to swim in the great northwest rivers, flourish for a while in the open ocean, and then swim up the rivers to lay their eggs and die.

I myself have not seen any of this, of course, but I am sure that it takes place.

Why do they bother, these salmon? No other fish requires such an elaborate apparatus of misery and migration. Perhaps the Pacific salmon was originally a freshwater fish separated somehow from its spawning grounds; perhaps they require fresh water to spawn. Perhaps the cycle is dictated by other factors. Yet here is the point that should provoke a doubt. The desirability, speaking fishwise, of an infinitely complicated reproductive routine is never demonstrated within Darwinian theory nor derived from general qualitative principles.

•••

I pass now to an account of certain modest triumphs of evolutionary thought. The subject at hand is monkey testicles; the question at large, why some of them are more substantial than others? Harcourt, Harvey, Larson and Short (a firm of accountants, actually), studying precisely this issue, reasoned that in polygamous primate systems in which each male competes with every other male for mating rights, those males with the largest, heaviest-hanging testicles should inevitably be favored in the competition. They do not say why. It follows that the ratio of testicle weight to body size should be quite different in chimpanzee societies, which are promiscuous, and gorilla societies, which are not. With careful restraint the authors note that "they have tested the hypothesis across a wide range of primates," although, for understandable reasons, they provide no data concerning the actual protocols of measurement. Their work has been discussed by Martin and May, who observe that the *general* relationship between polygamy and testicular weight, what Spanish-speaking researchers might call *cojonismo*, fails in the case of the polygamous squirrel monkey, whose breeding season is short but whose testicles are small when compared to the steadfastly monogamous cotton-top tamarin.

Having explored such facts in a human context only furtively, from the corner of a wet eye, so to speak, I cannot comment on the reliability of the research, but I am all for science, wherever it may lead – so long, of course, as I do not have to follow.

Checking Out

Every journey, I suppose, is an escape. I left for Paris on a dark northwest day in the fall of the year. Everything was dripping – the spiky evergreen leaves, the ends of staircase banisters, the trailing edge of bill-boards; even my breath entered the unmoving air as a humid puff and appeared for a moment to form a cloud. I clambered into the airport bus, bumping my head as I did, my suitcase banging irresolutely on the narrow stairs behind me. The black rubber mounting to my window formed a perfect frame: In the background, where Giotto would have painted a sunny mythical mountain, there was an unbroken ridge of wet green evergreens; in the middle, where Memling might have painted a turbid square, a dismal cluster of suburban shops – *Bruno's Pizza, Toys for Tots,* a restaurant inaccurately calling itself *The French Connection,* a massage parlor specializing in late-night commuters with fancy cravings; and in the near foreground, perfectly composed, a figure in a blue pea coat, a face that I knew as well as I knew my own, short black hair, tense eyes, the chin tilted just lightly, two mascara-streaked tears wandering over the cheeks.

The plane rumbled and groaned and clanked down the long runway at Seatac, and lumbered into the air. Below I could see the heavy cold waters of Puget Sound and the dense somber stands of spruce on Vashon Island. As the plane banked, and then turned steeply toward the north and east, it broke through the low-lying layer of cumulonimbus clouds that covers the northwest for much of the year, and there, impossibly bright, utterly clear, was Mount Rainier, sunlight on its snows, silent, sweeping, silvery, still, serene.

Some years before, I had spoken to Noam Chomsky about the theory of evolution. He told me to see Tom Bever at Columbia University, and then, almost as an afterthought, added that the man I really ought to talk to was Marcel Schutzenberger in Paris. Chomsky told me that while a Junior Fellow at Harvard, Bever had driven James Watson mad with his questions about the theory of evolution. It was an image I cherished until I actually met Bever.

I thought little about Schutzenberger until I published my first book; I sent him a copy with a hopeful note. Within days I received a handwritten response, thanking me for the book; days later, I received a second note. My little book, Schutzenberger wrote, was a "masterpiece of critical thinking." I had never been accorded praise on this order. I was deeply impressed.

My purpose now in traveling to Paris was ostensibly to discuss the theory of evolution with Schutzenberger. We had agreed by cable that he would arrange a hotel for me, something cheap.

I arrived in Paris that afternoon, and after a bouncing taxicab ride in a decrepit Citröen smelling vilely of Gauloises, I discovered that *Le Grand Hôtel de Paris* (three stars gleaming on the door, *bien sûr*) was located just off the Rue Monge, a street of positively supreme seediness, and that the hotel itself was small, dark, ugly, cramped, and battered. My room was dank. The wallpaper on both walls had peeled back, the better to accommodate a troop of bedbugs, who, in single file, were marching from left to right, arising from one crack in the wall to disappear after they had crossed the ceiling into another. There was a chipped porcelain washstand in the room, but no shower, and no toilet either, except for the low-lying, dust-filled bidet. When I sank onto the narrow bed, the thing wheezed melodiously, the mattress promptly curving into a parabola.

The monotonous hum of the jet exhaust was still very much in my ears; I looked up at the ceiling and endeavored to bunch up the curious tubular pillow that one finds in French hotels. I drifted off and actually slept for a few moments until roused unceremoniously by the hotel's clerk, who had, just an hour before, handled my American passport with the air of a man examining a dead fish, and was now standing outside my room knocking rhythmically on the door. I had a telephone call. It was Schutzenberger.

"You come to lunch, yes?" he said, jumbling tenses and moods in a single sentence, with only the charming imperative left intact. I said that I would.

Schutzenberger lived in the sixteenth *arrondissement*, in the sort of nondescript flat that in Paris signifies status but not necessarily wealth. The flat itself consisted of a series of large, sparsely furnished rooms; from somewhere, far away, the sound of chamber music drifted into the apartment. There were several grim pictures in the living room. In one, a stout, handsome woman stood staring somberly into space; in another, a small boy in shorts stood posed next to a drab horse, a brown, academic, nineteenth-century French landscape behind the two of them. The severe, iron-edged, double-doored windows were closed. It was very wam.

Schutzenberger was somewhat shorter than medium height, very thin, in his late fifties, I judged, with thick, long, grey hair, which flopped over his forehead and which he continually swept back over his head. At the door, where he greeted me shyly, he carried himself with severe erectness, as if his lower spine had been fused, and when we walked to the living room, I noticed that like many Frenchmen he held his elbows close to his sides.

"It is a *fantastic* pleasure to see you," he said in English, but with an accent so strong and so melodious that I thought for a moment that he was speaking French.

We sat in large, old, comfortable chairs and talked. After a time, Schutzenberger's Javanese wife, Hariati – a solid, immensely dignified, slow-moving woman with deep, wide, chocolate-brown eyes – came in with lunch.

When we had finished, we agreed to meet and talk again during the week. Our idea was to publish a paper together on the neo-Darwinian theory of evolution. When it was time for me to go, Schutzenberger drove me back to my hotel. He was an insanely wretched driver, and once went several blocks out of his way in order to pass a driver who had passed him first. "Oh, the *fuckair*," he said with satisfaction after we had overtaken the man and cut him off.

The Spaces of Life

The popular view of evolution tends to be a tight shot on a tame subject: The dinosaur, who did not make it; the shark, who did. In fact, the maturation of an organism is itself much like the evolution of a species. Only our intimate acquaintance with its precise and unhesitating character suggests, misleadingly, I think, that the two processes differ in degrees of freedom. For that matter, psychology, economics, political science, and history also describe processes that begin in a state of satisfying and undemanding simplicity and that end later with everything complex, unfathomable, chaotic. The contrast to physics is sobering.

The dynamics of evolutionary theories is often divided into two conceptual stages. In economics, there is macro- and microeconomic theory: the theory of aggregate demand, say, and the theory of the firm. Within linguistics, language is analyzed both at the continuous level of speech, and at some levels below, where things are discrete, a matter of the concatenation of words or morphemes. Biology, too, is double-tiered: Above, the organism prances; below, unseen, at a separate level entirely, its life is organized around the alphabetic nucleotides.

The disciplines of comparative anatomy and systematic zoology classify creatures into ever larger sets and sets of sets: There are individuals (dogs, say), species, genera, families, orders, classes, phyla, taxa, and kingdoms. The classification itself forms an intricate latticework, with individuals acting as the system's atoms. When it comes to classification, the obvious cases, of course, leap to the eye, but at the margins of the system, where the whale resides, difficult matters are decided by an appeal to different criteria – the whale is recognized as a mammal in virtue of the way it breathes and its method of bearing its young, but *not* its mode of locomotion.

By a *metric space*, mathematicians mean a collection of points or numbers over which a measure of distance has been defined. The integers constitute a metric space of an especially simple sort: The difference between any two numbers represents their distance. The classification that the biologist imposes on the natural world rests, in the end, on some felt (or imagined) perception of distance among biological creatures. This intuition is often reflected directly in the experience of biological sympathy. Toward the mammals we enjoy an especially warm feeling; not so the fish, still less the insects. The eyes of an ant, when magnified, represent only an alien form of life. The natural mathematical translate of sympathy is distance – a cold concept standing in for a warm one.

The very intricate classification achieved by systematic zoology commences at the level of the organism itself and works upward; wherever it goes, the classification organizes either individuals or sets of individuals or sets of sets of individuals, but *individuals*, in any case. In going dissectively downward, the biologist encounters objects of rather a different sort. DNA is a string whose elements are drawn from a four-letter alphabet; the proteins are strings again whose elements are drawn from a stock of twenty amino acids. As such, both strings belong to a wider family of string-like objects: computer programs, the sentences of a natural language, the elements of a formal system. DNA and the various proteins thus acquire by intellectual osmosis a very distinct conceptual and mathematical structure. Here, I daresay, biological intuition counts for very little; whatever classification the mathematician or biologist manages to impose on either DNA or the proteins he imposes by a form of groping in which *experience* (with similar objects) has come to replace intuition entirely.

An *alphabet* is a fixed and finite collection of elementary entities called words; the *universe of strings* over a finite alphabet, the set of all finite sequences whose elements lie in the alphabet itself. Now words, as I have suggested, are mathematical objects in their own right; indeed, along with sets, they constitute a great and noble class of non-numerical items in the mathematical universe. Under the usual – the standard – definition of distance in a space of words, the distance between two strings measures the number of separate changes required to bring them into perfect alignment. The metric space that results – the *natural* metric space – biologists often reject; indeed, *The Journal of Theoretical Biology* is riddled regularly with preposterous papers proposing some definition of distance based on a variety of strange biochemical properties. The theory of evolution implies, however, that change within biological strings comes about at random flash points where individual letters are scrambled. Some strings may change in an even more dramatic way, with whole blocks of letters being shifted at once. The least mechanism to which these operations may be resolved is the simple one of erasure and substitution – deletion and insertion. The elementary events of evolution at the molecular level thus lend to the natural metric a certain plausibility in the face of fancy competition.

It often happens, for reasons that are not mathematical at all, that a particular mapping between metric spaces is especially plausible. The English alphabet, for example, makes for two metric spaces: strings of letters, sets of words. Strings of letters are close if they agree in *spelling*; words, only if they agree in *meaning*. The natural mapping between these spaces simply serves to associate strings of letters with the words that they express. It is a peculiarity of this mapping that small typographic changes give rise to very large changes in meaning – "pant" to "part," for example, or "smell" to "smile." The metric spaces, I shall say, are not in *phase*.

What of the relationship between the alphabetic and biological spaces of life? The central dogma of molecular biology establishes that life in the large is a matter of a natural mapping between biological *strings* and biological *organisms*. This is a connection effected between two radically dissimilar worlds. There is thus no reason to expect these spaces to be in phase; the example of ordinary language might suggest that no one concept of distance covers both cases. In fact, the definition of distance in zoology owes nothing to the definition of distance in molecular biology. The two concepts cannot be expressed in a common vocabulary. These facts theoretical biologists often keep from one another, so that their implication, which issues when the facts are accidentally commingled, comes as an awkward surprise. Human and chimpanzee polypeptide sequences, for example, are, when considered simply as sequences, astonishingly alike; the degree of coincidence is close to 99%. And yet the species themselves, while close in some respects, are far further apart than a comparative analysis of their polypeptide chains might otherwise suggest. On mathematical grounds, this is just what one might expect. In theoretical biology, it is regarded as a paradox and is the occasion for a good deal of head scratching.

A Little Number on Complexity

Complexity and simplicty are metaphysical duals, like Yin and Yang, and, except for a vagrant connection to intuition, it hardly matters which is called what. Mathematicians and philosophers are interested in complexity for their own ends, yet in theoretical biology, and in ordinary life, the concept of complexity seems to have some *intrinsic* point or purpose. It is a simple fact, for example, that the human brain is more complex than the brain of a snail. But *complex* meaning what? And to what degree, if any?

Counting principles often seem as if they might provide some scheme for the measurement of complexity. Can we not say in general that the complexity of an object is ultimately a measure of the number of its parts? Or something?

This is not so much a definition as a suggestive idea. And one pointing in the wrong direction. Made of bricks, a heap and a house have precisely the same number of parts. But why count one (spatial) arrangement of a group of objects more complex than another, especially when all such spatial arrangements are arrangements of exactly the same objects?

Statistical mechanics suggests that the problem be resolved by conceptual ascent. The various states of a system of bricks (heaps or houses), the physicist would have us disregard. What counts is the way in which the states are *classified*. A house is intrinsically no more complex than a heap; but bricks wind up as heaps more often than they wind up as houses. It is heapness as a state that is simple (because common), and houseness as a state that is complex (because *un*common). Such is *complexity under a classification*.

Complexity *under* a classification is a concept that is obviously contingent *upon* a classification. In the case of houses and heaps, the heaps come first conceptually. These correspond to the various ways in which a group of objects might be rearranged. In sorting out the houses from the heaps, the philosopher, as well as anyone else, makes use of the fact that he simply knows what a house is. It is no easy task, however, to offer a precise account of this kind of knowledge, even in trivial cases. The most familiar of objects thus appear to resist a definition in anything like formal terms. This is a circumstance with which theorists of artificial intelligence are familiar. In organizing the micro-states of a physical system, the physicist need appeal only to a handful of obvious macroscopic parameters – pressure, volume, temperature, and the like. In sorting out heaps from houses, the philosopher ends – inevitably, I think – by appealing to a system of classification that is frankly Aristotelian. A house is a heap-like construction that satisfies certain human *ends* or *needs*. These are the organizing principles to which we appeal when we need to make ourselves understood.

When objects are classified under *these* concepts, they reveal their complexity with gratifying speed. A house, understood as an object with a complex function, is more complex than a heap, just because there are far fewer ways to realize a house than a heap. But isn't this just what the physicist might say? Not quite. His interests lie with states that are most likely to occur – *there*, entropy is at a maximum; my own, with states that are least likely to occur – *here*, entropy is at a minimum.

It is a curious fact that so many objects in the human and biological world turn out to be complex – curious, because only familiarity makes it expected, and a fact because it happens to be so. In any event, the concept of complexity under a classification is at a double distance from the concepts of physics. It is a notion applicble to systems not obviously at equilibrium, and it makes use of a system of classification that is not found in, and cannot be reduced to, physics at all.

The Language of Life

When it comes to language, there is syntax and semantics. Phonetics is the province of the specialist; pragmatics remains a pale, Albino dwarf. To semantics belongs the concept of *meaning*; to syntax, the concept of *grammar*. Language-like systems go beyond the natural languages, of course. There are computer programs, algorithms generally, formal systems, purely algebraic objects such as semi-groups, and even, perhaps, the sequences of nucleotides and amino acids.

The construction of strings within a language-like system is a matter achieved by *concatenation;* most (but not all) language-like systems are large in the sense that concatenation may be iterated at will. Meditating on the matter in the late 1950s, and regularly thereafter, Noam Chomsky argued that every natural language is infinite by virtue of its recursive mechanisms – conjunction and alternation, for example – and, simultaneously, that such mechanisms are recursive by virtue of the fact that every natural language is infinite. Whatever the truth of this claim, the language-like systems, *if* they are infinite, are countably infinite and no bigger. They may be put into one-to-one correspondence with the natural numbers.

Going further toward a definition of a language-like system involves an excursion into the badlands beyond triviality. Linguistics, Maurice Gross once remarked to me (in the fall of another year, in Florence), admits of but a single class of crucial experiments. Native speakers of a given language are able to determine whether a given sentence is grammatical. Experiments of this sort exist because no natural language encompasses the *whole* of a set of strings drawn on a finite alphabet. This is a curious and interesting *fact* – a principle of *fastidiousness* – and one that the sheer concept of communication might not otherwise suggest.

Language-like systems, then, admit of a primitive distinction between grammatical and ungrammatical strings. What of their complexity? Thus far, complexity has emerged in these pages as a Siamesed concept. There is Kolmogorov complexity – the first head – and the concept of complexity under a classification – the second. The principle of fastidiousness might suggest that the grammatical strings of a natural language are highly complex under the obvious classifications imposed by grammar. And, indeed, they are. It is characteristic of language-like systems that they give rise to a very *large* number of combinatorial objects and a very *small* number of meaningful or grammatical combinations. On the other hand, those same strings are generally *low* in Kolmogorov complexity. Natural languages are redundant. (So for that matter is the mammalian nervous system – another queer fact.) This implies, rather unnervingly, that one and the same object is both simple and complex. Just so. But the sense of complexity is different in each case.

Weak Theories

The vitalist believes that life cannot be explained in terms of physics or chemistry. In nineteenth-century Germany or France, at least, his was the dominant voice before Darwin. Natural philosophers, such as Cuivier or von Baer or Geoffroy Saint Hilaire, dismissed mechanism with troubled confidence. Orthodoxies have subsequently reversed themselves with no real gain in credibility. David Hull, surveying this issue, concludes that neither mechanism nor vitalism is entirely plausible, given the uninspiring precision with which each position is usually cast. I agree. To the extent that the refutation of vitalism involves the reduction of biological to physical reasoning, the effort involved appears to me misguided, and reflects a discreditable, almost Oriental, desire for the unity of opposites. Our intellectual experience *is* sharply divided by subject. Each science extends sideways and then simply stops. The philosopher with no vested interest in doctrine of dogma might reasonably conclude that the various sciences are controlled by a committee of Gods, with mathematics under the influence of an austere Artin-like figure, and biology directed by a deity that is furious, bluff, subtle, devious, and illiterate.

Still, the philosopher of science is bound to wonder why so many other philosophers have in the end remained partial to the reductionist vision and hence to mechanistic thought in biology. David Armstrong, J. J. C. Smart, Michael Ruse, and even the usually cagey W. V. O. Quine, call on elegance to explain their attachments. Were the sciences irreducibly stratified, one set of laws would cover physics, another biology, and still a third, economics and urban affairs, with the whole business resembling nothing so much as a parfait in several lurid and violently clashing colors. This is an aesthetic argument, and none the worse for that, but, surely, none the better either.

If elegance is inadequate as a motive for mechanistic thought, intellectual anxiety, realized unconsciously, is not. Vitalism commences from the conviction that nothing in our experience is much like the life that ripples and bubbles around us. It thus represents the first and most obvious of principles. Now mathematical physics is not only the preeminent discipline of our time – it is where the laws are. Evolutionary theories in biology are *weak* in the sense that they are not directly sustained by the laws of physics, and, worse, weaker still in being curiously counter-physical. They thus remain surprisingly resistant to confirmation. Were biology, like chemistry, truly an aspect merely of physics, the sceptic would get short shrift. The answer in physics to whether what works, works, is simply that is does. On observing his distance from mechanistic *theories*, the modern biologist thus tends vigorously to affirm his devotion to mechanistic *thought*, an interesting example of wish fulfillment in intellectual life.

With this observation I conclude the diagnostic portion of my deliberations. What follows is therapeutic. Scientific theories, it seems to me, are by their nature intended to be general. When I speak of a theory, I follow the logician's lead: A theory consists of a consistent set of sentences in a given language. The set-theoretic or algebraic structures in which they are satisfied comprise their models. Two models that share precisely the same structure are *isomorphic* and hence elementarily equivalent in the sense that they satisfy precisely the same sets of sentences. Isomorphism is an indication of indifference. What I am after is a weaker notion entirely – *partial* similarity in structure. I know of no way, unfortunately, to define this concept precisely; I am reduced to offering examples. The various kinds of alphabetic objects – computer programs, algorithms, the elements of a natural language, proteins – seem to me, when recast abstractly, to form an overlapping series of models each of which shares with its neighbor some essential sameness in structure. Whatever the details, similarity in structure is bound to be a matter of degree, so that it makes sense of sorts to say of two models that they are at a certain distance from each other. In a context such as this, a principle of *congeniality* would appear to hold. Suppose I say that a theory satisfied in a certain model M is *general* just to the extent that it is satisfied simultaneously in all models within a fixed distance of M itself. Generality in this sense is a form of stability and brings to mind Dr. Johnson's admonition that the soul must ultimately repose in the stability of truth. The principle of congeniality now emerges as a corollary. A theory is plausible, I shall argue, just to the extent that it *is* general.

To see an analogy between the operations of life, on the one hand, and the operations of language, on the other, is to raise the question of whether the laws of biology have a natural and legitimate, if partial and inconclusive, interpretation in linguistic terms. I myself am indifferent to the fate of the Darwinian theory and perfectly prepared to believe, along with Wickramasinghe and the luckless Hoyle, that life originated in outer space or that the Universe-as-a-Whole is alive and breathing stertorously. Yet if Darwinian theories work in life, they should work elsewhere – in language-like systems, I should think. Should they fail there, this may be taken as evidence for the inadequacy of Dawinian theories; it may also mean that the analogy is itself misconceived. Yet if the analogy *is* misconceived, life stands revealed as a phenomenon much unlike anything else in our conceptual experience. One cannot help wondering whether, and in what terms, a process so unique and so singular might be explained at all.

I stress this point only because it has so often been misunderstood.

Language-Like

It was John von Neumann who gave to the idea that life is like language a part of its curious current cachet. The last years of his life he devoted to a description and orchestration of a variety of cellular automata, showing, in a partial fashion, that when properly programmed these abstract creatures might carry on a variety of lifelike tasks – reproduction, for example. Some years before, McCulloch and Pitts had constructed a series of neural nets in order to simulate simple reflex action. The American logician Kleene demonstrated that the nets had the power of finite-state automata, and were capable of realizing the class of regular events. Von Neumann's automata had the full power of Turing machines. Michael Arbib, E. F. Codd, G. T. Herman, A. Lindenmayer, and many others, have carried this work forward, with results that asymptotically approach complete and utter irrelevance.

Yet the analogy between living systems and living languages has not lost any of its brassy charm; indeed, I favor it myself. Unlike an argument, however, an analogy stands or falls in point of plausibility. Good arguments in favor of bad analogies are infinitely less persuasive than bad arguments in favor of good analogies. One of the joys of analogical reasoning is the vagueness with which the line between success and failure may be drawn. Certainly the proteins, to stick with one class of chemicals, may be decomposed to a finite base – the twenty amino acids. The precise and delicate, dance-like steps that are involved in their formation suggest, moreover, that they satisfy some operation abstract as concatenation. On the other hand, the number of proteins, although large, is finite. A natural language contains infinitely many sentences. The score in favor of the analogy between life and language is thus even.

The grammatical strings of a language-like system, I have argued, are low in Kolmogorov complexity and strongly non-random. What of the proteins? In *What is Life?* Erwin Schrödinger argued that living systems must have recourse to what he dubbed an "aperiodic crystal" in order to store information. Crystals are repetitive in their structure and poor in information. There is a splendid effulgence to the vocabulary of theoretical biology that it would be uncharitable to ignore. H. P. Yockey identifies the order of a biological system with Kolmogorov complexity; so does R. M. Thompson, a mathematician who in writing about theoretical biology alternates between information theory and a pious endeavor to communicate to the reader his appreciation for the many faces of Krishna. On the other hand, G. J. Chaitin and R. M. Bennett identify biological order with algorithmic *simplicity*. A division of intuition on so fundamental a point may suggest a degree of conceptual confusion approaching the schizophrenic. Jacques Monod, whose metaphysical attitude toward molecular biology was one of defiant chirpiness, drew especial attention to the random character of the polypeptides in *Le hasard et la necessité*, a book that to my mind represents one of the great embarrassments of contempory French science. His arguments have been accepted by many molecular biologists and philosophers. In fact, the evidence leading to his conclusion is fragmentary; the standards of randomness to which he appealed, imprecise. Thus it struck Monod that in knowing, say, 249 amino-acid residues in a chain 250 residues in length, one could yet not predict the last member of the chain. Much the same is true of the simple sentences in English, of course.

Nonetheless, I am in sympathy with Monod to this extent. It is unlikely that the analogy between life and language will be profitably pursued on the atomistic level of the nucleic acids or the polypeptides.

In the Sixteenth Arrondissement

Our routine was always the same. We would meet in the late afternoon in Schutzenberger's flat. There would be nuts or fruit on the table. Schutzenberger would be wearing a gray-green military shirt, green tie, pants with suspenders, socks and no shoes. He looked very elegant, the creases in his clean clothes matching somehow the creases in his face, as if the two of them, those clothes and that face, had been through a difficult campaign together, as, indeed, I suppose they had. I brought a notebook to our meeting with the idea of keeping a record of our conversations. I sat facing Schutzenberger in an oversized stuffed armchair. Schutzenberger was curled on the couch.

Bantering or badinage was out of the question, really; Schutzenberger needed to speak in blocks and resented interruptions. He had a wonderful passion for his own ideas. I would always try to begin things formally with an account of what we had done, where we were. It was a waste of time. Schutzenberger had no use for the continuities of argument. What he emphatically embraced yesterday, he indignantly rejected today, or simply washed away with a wave of his hand. He needed in some fundamental way to feel himself spontaneous, recreating, if necessary, arguments that he had already established; years later I found that he remembered the twists and turns of our discussions almost as if he saw them inscribed on a road map.

The chicken scrawl of my notes, now that I read them over, makes mention of stability, probability, statistics (which Schutzenberger pronounced a graveyard), ergodic theory, the Kolmogorov theory of complexity, geology, the theory of plate tectonics, a certain problem in elementary physics mentioned by Mach, mathematical physics itself (a subject, Schutzenberger said, in which he could spot the mathematics but not the physics), the theory of evolution (our ostensible topic), finite state automata, the algebraic theory of non-commutative semigroups, Lie Groups, the theory of representations, the system of pronouns in French, modern taxonomy and systematics, natural history, the social life of the chimpanzee, the irrelevance of Freudian theory to French psychiatry, and the nature of the Soviet secret police. Like many brilliant talkers, Schutzenberger's great need finally was to talk about himself, and my great need was to listen to him talk. He had a marvelous capacity for malicious contempt. A favorite target was Ilya Prigogine, but Schutzenberger had something evil to say of everyone: H. A. Simon, whose Nobel Prize Schutzenberger took as the fulfillment of a vindictive Biblical prophecy, Norbert Wiener, whose name he pronounced Nor*bairt*, Patrick Suppes (the Soup), Paul Samuelson, Dana Scott, B. F. Skinner, Albert Einstein (whom he cheerfully dismissed one day as an old fucker), and a certain, very prominent biochemist from California, who had come to Paris to lecture at the *Collège de France*, and who had flabbergasted his hosts by asking that the *Parc Monceau* be opened at six in the morning so that he could jog through its walkways. Like Evelyn Waugh, whom he deeply relished ("Is he not delicious," he remarked to me once), Schutzenberger formed his political and social principles chiefly on the basis of the degree to which they placed him in the minority, and so expressed at every opportunity his dislike for every conceivable ethnic, sexual or racial group; but all this was done with such high humor and obvious relish, and directed at such a kaleidoscope of shifting targets, that no one listening could have possibly taken offense.

Later he would wind down, his mood turning elegiac. He would talk about the past. Schutzenberger had grown to young manhood in Paris during the war, not a happy experience, but one with which he identified utterly. Once he mentioned the valley of the Loire, where he had spent summers as a child. The image of that long, gentle river slipping between heavy, lush fields, he said, would be with him forever, a remembered thing, his to consume for as long as he lived.

Arguments Gone Bad

The theory of evolution is haunted by an image and an observation: The first, that of the hapless chimpanzee, typewriter-bound, endeavoring quite by chance to strike off the first twenty lines of Hamlet's soliloquy; the second, the comment of a Jansenist logician, who remarked, quite sensibly, that "it would be sheer folly to bet even ten coppers against ten thousand gold pieces that a child arranging at random a printer's supply of letters would compose the first twenty lines of Vergil's *Aeneid*." Image and observation do not quite cohere into a single argument. It is clear in neither case *how* the imagined stochastic experiment is to stop. Still, the notion of randomness yet dominates evolutionary thought, and there it sits, toad-like and croaking.

On the simplest and most intuitive conception of probability, what can occur is weighted against the background of what might occur: the flush of five diamonds, all other combinations of the cards. In poker, there are 2,598,960 five-card hands, but only 5148 flushes. It is their ratio – .002, as it happens – that one might expect to observe if flushes are being counted over the longest of long runs.

The evolution of life on this planet is, as Darwin realized to his own astonishment, not a hurried affair. Early on, Darwinian biologists got rid of the theological limits set to the age of the earth by Bishop Ussher and others in the seventeenth century; the scale within which Darwinian evolution might have worked is bounded by something like five billion years. Nineteenth-century biologists assumed that whatever else one might say about Darwinian theories, they would not fail for lack of time. This thesis twentieth-century biologists have carried over intact.

Five billion years is apt to seem long if one is counting the minutes; it is not long enough to sample on a point-by-point basis a space whose cardinality is roughly 10^{15} – touching base with a new point at every second, say. And yet there are 20^{250} possible proteins in nature. This is a number larger by far than the expected life of the universe measured in seconds. In a space of this size, the odds against discovering any specific protein strictly by chance are prohibitive – 1 in 20^{250}.

According to the central dogma, DNA is a molecule with a double life. It exists in the first place to exist once again; and it exists in the second place to express precisely the genetic identity of the creature in which it is embedded. It is thus poised between the concepts of immortality and identity and so occupies a transcendental perch in the scheme of things. Yet this is too static an image. Chemical processes of the most exquisite refinement take place in the bacterial cell and wind their way through a series of graceful metabolic loops. However much they dream of remaining biological big shots in the far future, the nucleotides manage to keep things humming in the unavoidable present. Their activities take place with speed, precision, and an absolute economy of effect.

Certain genes within the bacterial cell, Murray Eden observes (I shall bring Murray to life later), "are organized into larger units under the control of an operator, with the genes linearly arranged in the order in which the enzymes to which they give rise are utilized in a particular metabolic pathway." Thus suppose that the bacterial cell needs to convert *A* into *B* and *B* into *C*. The nucleotides are so arranged in the bacterial cell that the order in which *they* are read corresponds to the order in which the proteins pass through a metabolic pathway. The particular elements of the gene that initiate and then stop this zipper-like effect are known, respectively, as *inducers* and *repressors*. An *operon* consists of a set of individual genes, however long, under the control of repressors and inducers. It is the operon that determines a particular sequence of chemical reactions within the bacterial cell.

I arrive now at the point to this exercise. "What is the probability," Eden asks, "that a arrangement of unordered genes will organize certain groups into operon clusters?"

Low, one would imagine, and this on general probabilistic grounds. What comes as a surprise is a matter of degree. "One would need an average population of bacteria," Eden writes, "of 10^{30} (about 10^{13} tons, or a layer on the surface of the earth two centimeters thick) if one expected to find a single ordered gene pair in five billion years."

•••

"Biologists," Peter Medawar has observed, "in certain moods are apt to say that organisms are madly improbable objects or that evolution is a device for generating high degrees of improbability. I am uneasy about this entire line of thought," he continues, "for the following reason:

> Everyone will concede that in the games of whist or bridge any one particular hand is just as unlikely to turn up as any other. If I pick up and inspect a particular hand and then declare myself utterly amazed that such a hand should have been dealt to me, considering the fantastic odds against it, I should be told by those who have steeped themselves in mathematical reasoning that its probability cannot be measured retrospectively, but only against a prior expectation. . . . For much the same reason, it seems to me profitless to speak of natural selection's generating improbability, . . . it is silly to be thunderstruck by the evolution of organ *A* if we should have been just as thunderstruck by a turn of events that had led to the evolution of *B* or *C* instead.

Medawar is roughly right about probability. The fallacy to which he refers is the error of *retrospective specification*, and consists precisely in reading back into a sample space information revealed only on the realization of a particular event. In poker, as I have mentioned, a deal distributes *n* hands of equal probability: 1 in 2,598,960. This sample space is specified retrospectively if one hand in particular is contrasted with the 2,598,959 hands that remain, and probabilities assigned to the partition thus created. What appears initially as one among equiprobable events becomes under retrospective specification an improbable event in a sample space of only two points. It is embarrassing for an author to point out such things.

Still, Medawar is wrong in the general conclusions he draws from this paragraph. When it comes to poker, neither the cardsharp nor the statistician is much interested in the set of all five-card sequences. In playing the game, the cardsharp partitions the various hands into equivalence classes of uneven size. There is the royal straight flush, to begin with, and of these there are only four. Next comes the happy class of straight flushes; after that, the flushes themselves, and then the four of a kinds, and beyond that, the full houses, the straights, three of a kinds, pairs of pairs, pairs, and, finally, whatever is left. It is into the whatever-is-left category that the majority of hands actually fall. It is quite true, as I shall now say for the third time, that the full set of five-card sequences in poker comprises a uniform sample space in the simple sense that each particular sequence – considered simply *as* a sequence – receives precisely the same probability. The statistician's eye, however, moves naturally upward – to the *classification* of five-card sequences. The elements of this classification he counts as a sample space in its own right. Sitting at a green, felt-covered table in Las Vegas, a sun-visor shielding his eyes, he – that cardsharp turned statistician – counts himself *lucky* when, of the five cards passed his way, four turn out to be aces. His sense of dawning delight is reflected in the odds, of course. There are only a few ways in which five cards might yield four cards of the same suit; very many more ways in which the cards might turn up blank. The odds associated with the various partitions in poker are specified in advance; there is nothing backward looking here.

Medawar's argument, on its face, thus involves rather an uninspiring confusion. Needless to say, a separate argument is required to show that the confusion extends to evolutionary thought. In this regard, consider the human eye, and suppose that the eye has been resolved to its molecular constituents – a very large class of proteins, say. The living human eye, Medawar might argue, represents one arrangement of its constituents. Any other might have done as well. In admiring the structure that results, we suffer from misplaced awe, like a toad contemplating a dog.

Does this argument carry conviction eyewise? Is it reasonable to suppose that any other arrangement of the eye's constituents would result in an eye? In anything at all? To ask this question is to repair once again to the concept of complexity under a classification. The argument is now moving in all the old, familiar circles. An eye is an organ that achieves its identity by means of its *function*. Eyelike configurations, and eyes, thus admit of a second-order classification, one that goes beyond any molecular considerations to appeal to the constraints and categories by which the living world is organized. As in the case of cards, one sample space gives way to another. Now the analogy between life and language that I have been pursuing these many pages suggests that life, like language, is complex under many of the obvious systems for its classification. Complexity thus interpreted carries an obvious numerical implication. When set against all other configurations of its constituents, the living human eye is *singular*. (That sense of awe I rhetorically dismissed, I now embrace.) This is an observation that might well have been prompted by common sense or the experience of surgery. The operating-room bungler, slicing into the wrong artery with a great gust of casual enthusiasm, is hardly likely to *improve* the performance of an organ – the kidney, say. How else to explain this fact than by assuming that biological structures are complex under their various functional classifications? But the analogy between life and language gains its point and its purity only when alphabetic structures are compared – nucleotides and words, say. In discussing the eye, I have appealed to a certain analogical spillover.

Meaning

Linguistics, I have observed, is possible if only because human beings have strong and reliable intuitions about natural languages. The polypeptides are alien strings, accessible only through an arduous act of the biochemical imagination. The concepts of grammar affect a segregation of strings; beyond grammar, aloof, untouchable, there is *meaning*. The two concepts do not coincide. Some grammatical strings – in a natural language, at least – are grammatical and meaningless; others, meaningful but ungrammatical. Yet meaning and grammar belong together, yoked pairs in the same corner of some dimly understood conceptual space. An algebraic system of strings in which no distinction of grammar or meaning is recognized is profligate, and pointless (for purposes of communication) because of its profligacy.

In a pre-analytic sense, the concept of meaning indicates a kind of coherence and admits of useful applications in domains quite beyond the language-like systems. A life well spent is meaningful: Its parts form a pattern. Filled with life, biological creatures are full of meaning, a kind of blunt, indisputable purpose. In death, this meaning disappears, and what is left, the corpse and its grim constituents, seems suddenly to lose the integrity of the creature itself and becomes, instead, a thing among other things, an object merely.

To the vitalist, living creatures instantiate some unique property that remains unseen elsewhere – in the domain of objects studied by mathematical physics, for example. In death this property vanishes, like a fluid evaporating. In mechanistic thought, the passage from life to death is rather like a phase transition, a singularity in the trajectory of the organism that reflects, as it must, *only* a change in its constituents, a variation in the underlying pattern of construction.

Let me pursue for the moment the idea that a biological creature represents one particular configuration of its basic elements. These basic elements I identify with the proteins. Now the proteins, I am supposing, may, like letters, be arranged in any number of different ways. The unalterable fact that living systems die indicates that some of these alternatives fail to preserve life and hence meaning. A complex organ within a complex organism is rather like a jigsaw puzzle. With even a single piece or a single protein out of place, the picture remains incomplete, the organ inadequate. No less appears true of the whole in which the part is embedded. The central dogma of molecular biology establishes a relationship between strings of nucleotides and strings of proteins. To the extent that the whole of a biological organism may be resolved to its protein-like parts, the central dogma establishes a more indirect relationship between molecular biological order and order or meaning in the larger sense of life. This relationship has an inverse. If only certain forms of life have meaning, then this too is reflected, as it must be, in the universe of molecular biological strings. If certain protein ensembles are meaningful, and not others, this suggests, but does not imply, that the same distinction is palpable on the level of the individual proteins. The term *viable* I mean as a biological coordinate to the concepts of meaning and grammar. A protein, I shall say, is viable just in case it achieves a certain minimum level of biological organization and usefulness. What level? What kind of organi-

zation? Usefulness in what respect and to what degree?
I have no idea.

•••

Like the restless eye that must move in order to see, our intelligence comes to judgment comparatively, weighing what we know of one theory in terms first made comprehensible in another. Nothing is quite like life except life itself. Still, certain systems are life-like in at least some of their aspects. From the point of view of evolutionary thought, life on the level of the nucleic acids is combinatorial, and the mix of letters and messages involved in the transmission and replication of information calls to mind both computer programs and natural languages.

To place life in the context of the language-like systems is to invite what psychologists refer to as conceptual discord – the experience a man has in explaining his mistress to his wife and his wife to his mother. A natural language is a system strongly infused with meaning. The distinction between sentences that are meaningful and sentences that are meaningless has thus an absolute character. The speaking-steps that we take move us inevitably and unconsciously from meaning to meaning, rather as if we were all tapping our way along a series of rocks perched precariously over an inky and meaningless void (as, indeed, we are, although just how those rocks remain suspended is a very great mystery). This point may now be followed by a principle: Meaning and randomness are opposites. The sentences inscribed on the printed page, when altered arbitrarily, give up their meaning with a gasp and soon thereafter transform themselves into a sullen string of symbols. They are then no use to themselves or anybody else. Thus if living systems *are* language-like, random changes should be similarly sense-destroying. (There is no point to an analogy if the thing cannot be exploited.) Yet random changes are precisely the means and the mechanism by which the Darwinian theory achieves its effects. Over time, the printed page of life should acquire something like a bug-spattered look. In fact, living systems seem cheerfully indifferent to a random permutation of their genes and carry on as if they had some secret mechanism guaranteeing that it all comes right.

This is an observation first made by M. P. Schutzenberger. "If we try to simulate such a situation [i. e., life] on the computer," he remarked at the 1966 Wistar Symposium, "we find that we have no chance even to see what the modified program would compute; it just jams." Biologists, hearing Schutzenberger out with their usual attitude of anxiety and aggressiveness, misunderstood the point entirely. Richard Lewontin, for example, – strange how many of these sad sacks come from Harvard – insisted perversely that whatever works in life, works, a thesis not under dispute. "Can I give you a practical experience," he argued in full voice, "where there is no jamming and no loss of meaning?" That practical experience turned out to be a description of the formation of tryptophane synthetase. Patiently Schutzenberger explained to Lewontin that what was at issue was not *whether* life works, but *how* it works, another matter entirely.

As the discussion proceeds – these verbatim records are a source of great amusement – everyone becomes progressively more befuddled; from time to time, Schutzenberger managed to swirl his cape to fact the panting Lewontin, but the artistic effect is spoiled for the reader by the peevishness of the exchanges.

Pedagogical Impact

In the fall I received an invitation to lecture at the Center for Theoretical Physics in Trieste. I stopped in Paris to visit Maurice Gross. When he picked me up from my hotel, he mentioned that Schutzenberger, having arrived moments before from Brazil, would join us for dinner. We ate in a stiff, solemn little restaurant tucked into one of a series of arches, somewhere near the Opéra and just off the Rue Saint-Honoré. I ordered kidneys by mistake. They arrived at the table swimming in some wet, red, faintly acrid sauce. I looked down at the poor limp little things with a marked lack of enthusiasm. Schutzenberger observed my discomfort and promptly undertook an analysis of my culinary provinciality for the benefit of the company. He had, however, ordered the fish himself, and in any case ate very little. Gross eyed my plate with the air of a man saddened by waste. Afterward, we ambled down the Rue Saint-Honoré, past those wonderful shops, which even at night glowed with efflorescence of wealth, and caught up on gossip. How was I doing? About to be fired. Again? Yet again. Had I performed an utterly unnecessary act of personal conscience? Was I skipping my own classes? Or was I involved in some *schweinerei* with my students? Or was it, Schutz suggested, a combination of the three? Hearing that it was, indeed, a combination, with several items yet unrecorded, Schutz decided to cheer me up by describing his own theory of pedagogical impact. At Poitiers, where he had taught before coming to Paris, he had scheduled all of his classes in mathematics during the hours between two and four in the morning. The great solemnity of the hour, he maintained, maximized the impact of his teaching.

"Did anyone ever show up?" I asked.

"Of course not," said Schutz briskly. "Me neither."

It was then that Schutz told me that he and Gross had arranged for me to spend a year as a professor of mathematics at the University of Paris. They thought that a year would give me time to breathe.

I needed all the air that I could get, and wrung both their hands in gratitude.

There to Here

Life, as I have said, loiters over two metric spaces. The first is alphabetic, the second, zoological. Evolution involves a drama in the large, at the zoological level, where various creatures live and die, but the changes effected there (opposable thumbs and the like) reflect the accidental changes in the nucleotides. The spaces are thus indirectly correlated.

In evolution at the molecular level, one amino acid is dropped from a protein string, another is inserted: *Make way, Move over, Get Out, Get Lost*. Even if the biochemical process is more complicated, as no doubt it is, it may mathematically be resolved to a series of discrete and finite steps. Whatever the details, to the extent that life changes, the proteins that animate life do so as well. These changes represent a *path* in the space of all possible proteins. It is on the level of paths, and not proteins, that the analogy between life and language may best be exploited.

The child and the chimpanzee, confronting the unyielding printer's blocks or the typewriter, undertake their literary efforts by means of a mechanism that is entirely random. Setting out, the chimpanzee is no more likely to strike T than any other letter. The odds in his favor are 1 in 26 – not counting spaces. Each effort in his enterprise is independent; the odds at each trial do not change. T having been struck, the odds in favor of striking O remain at 1 in 26. The odds in favor of first hitting T and then hitting O, however, are 1 in 26 multiplied by 1 in 26 – 1 in 676, a larger number entirely. In a general way, the odds against forming a particular word rise dramatically in proportion to its length.

Supposing letters varied at will, how many steps does it take to go from "choice" to "chance"? Two at a minimum, is the brief, buoyant response. And if the changes are perfectly random? Or if the goal of this exercise is to start at "choice" and stop at *any* meaningful word?

Despite the fact that I do not know the answers, I am prepared to admit that these are not trivial questions. Our sense of things in this context (a world of words) depends, I think, on the contrast between all six-letter words and those that happen to make sense. Two mathematical spaces are thus at work, playing over one and the same set of combinatorial objects. The first space consists of words understood typographically; the second, of words infused with their meaning and distinguished not by their shape but by their interpretation. These two spaces are plainly *not* in phase.

In explaining his own work, George Pieczenik told me to consult a paper by Murray Eden. It was a name I did not know. Eden had published his article, "Inadequacies of Neo-Darwinian Evolution as a Scientific Theory" – rather a heavy-handed title – in the *Proceedings of the Wistar Symposium*, and no sooner had Pieczenik mentioned the thing to me than I began encountering referenced to it in various places. How very odd that what one comes to know all of a sudden seems well known. Years later I met Eden at a dinner party given by Maurice Gross in Paris, another point of confirmation in a universal law of familiarity. He was modest and courteous, perhaps a little less than medium height, in late middle age, with a round face and a high forehead and demoralized eyes. He had six children. His wife was attempting to learn French with no great success, suggesting to me over dinner that the trouble was that the French took perverse pleasure in speaking rapidly. Almost everyone at the table had read and admired Eden's paper; even those academics uninterested in biology had the sense that it represented a significant achievement. Without quite being asked, Eden suddenly remarked: "It took everything I had to write that thing." I found this a wonderfully touching admission.

Hey Murray! Je ne vous ai jamais revu et quoqu'il y ait bien peu de chance que vous voyiez mon livre, permettez-moi de vous dire que je vous serre la main bein cordialement. J'espère que votre femme aura meillure chance avec la langue française.

Murray Eden begins his difficult paper by computing the size of the space of all possible proteins. That number I already know: 20^{250}. (There are twenty separate amino acids; each protein is roughly 250 residues in length.) Eden then compares this space with the space of all proteins that have at one time or another actually existed on the earth – in whatever form. The size of this space he fixes at 20^{52}. This is a much, much smaller number. (Inevitably, an element of fantasy enters into calculations of this sort.) These spaces may be analogically coordinated with words. The first space corresponds to the space of all possible words of a fixed and given length; the second, to a subset of this space. Between these spaces is all the difference between what might be and what is. In considering the play between the spaces, Eden remarks, "two hypotheses suggest themselves:

> Either functionally useful proteins are very common in this space [i. e., the space of all possible proteins], so that almost any polypeptide one is likely to find has a useful function to perform, or else the topology appropriate to this protein space is an important feature of the exploration: that is, there exist certain strong regularities for finding paths through this space.

The argument carries over to words. The *meaningful* English words, for example, makes up far more than a mere list. Certain sound combinations do not occur in spoken or written English – lavish consonant clusters, for example. The distribution of meaning within the space of all six-letter words is governed by strong phonetic regularities. This suggests, provisionally, that the meaningful English words are *homogeneous*. Any two words drawn from this portion of the space are apt to resemble one another, like fish drawn from the same portion of the pond.

In asking whether functionally useful proteins are common in the space of all possible proteins, Eden is asking, in effect, whether *any* combination of amino acids constitutes a suitable protein, or whether only *some* do. This is an interesting and perceptive question. I believe that Eden was the first to ask it, although it is a question that might occur naturally to a mathematician.

When it comes to the proteins, then, there are just two possibilities. Either anything goes, and hence anything counts as a possible protein, or some choices are marked out of bounds, just as certain word-like sequences in English do not count as possible English words. "We cannot now discard the first hypothesis," Eden writes, "but there is certain evidence which appears to be against it: If all polypeptide chains were useful proteins, we would expect that existing proteins would exhibit very different distributions of amino acids." Eden's example involves the alpha and beta chains of human hemoglobin. One form of hemoglobin has 146 amino-acid residues; the other, 140. The two chains may be set down, side by side, and matched, residue by residue. They agree at sixty-one points; there are seventy-six points at which they differ, and nine points at which no match is possible because the chains are not of the same length. It is plausible that one chain was derived from the other, or that both were derived from a common ancestor. What is curious about these pairs of proteins, however, is the fact that even though the chains do not agree completely in the order of their amino acids, they do agree in their *distribution*.

This is reason enough, Eden argues, to suppose that the proteins themselves have been drawn from a homogeneous sample.

Precisely the same observation may yet again be made on an imaginary population of words. If any combination of six letters could count as a possible English word, then it would be surprising to discover strong regularities in the words found in a dictionary. (In fact, the distribution of letters is similar in most English words.)

In the case of English words, some paths lead directly from one word to another – "choice" to "chance" by a path of two steps. In the case of life, Eden observes, "the actual path lengths are limited by the number of generations in the organism's history, so that the long paths are inaccessible, only the short ones have been taken."

Now, natural selection, on the neo-Darwinian scheme, is what mathematicians call a local operation. Each change in a population of proteins is evaluated in terms of the benefits that the change *immediately* confers on the organism. What is forbidden by the Darwinian theory is a definition of fitness pitched forward. Natural selection is not an operation that works by anticipation. In moving from "choice" to "chance" in just two steps, the mathematician achieves a certain economy of effect only because he knows precisely where he is going. Were natural selection at work, "chance" would not figure as a target of the operation; the mathematician who starts at "chance" and simply hopes for the best cannot reasonably expect to arrive at "choice" by the shortest or most direct route. And so, too, with hemoglobin.

The short path between proteins not only preserves meaning, it is significant because it is short. But while the short path in Eden's example is shorter than all the rest, it is still long – 2,700,000 generations – and this only to change a single protein. Other paths, by definition, are far longer.

A double dilemma, then. What, if anything, inclines

a molecular biological system along any meaning-preserving path; how is the shortest path ever found?

The Weizenbaum Experiment

The idea that some definition of meaning is stamped directly onto the nucleotides or the amino acids of a living system is neither supported nor contravened by neo-Darwinian doctrine. On these issues that theory is silent. Yet some evidence, while difficult to interpret, does seem to show that complex proteins are astonishingly low in specificity – high in Kolmogorov complexity; and *simple* under any number of relevant classifications. In an article in *Science*, for example, a quartet of molecular biologists (C. Kaiser, D. Press, P. Grisafi, and D. Botstein, 1987) offer a report on their own recent work. Their ambition is to discover the degree to which a specific protein sequence is resistant to random changes; their ingenious idea, simply to insert random sequences within the protein and then measure the extent to which the protein is yet able to carry out its biological functions. To their astonishment, and mine, it goes without saying, about one fifth of these essentially randomized sequences carried on with no troubling sense that some semantic catastrophe had overtaken the polypeptides. But evidence of this sort suggests the case in which a man is told that he does not suffer from one disease while learning simultaneously that he suffers from another.

We were sitting together, Schutzenberger and I, noodling and doodling. As so often happens in conversations of this sort, we were very much engaged in chasing down a number of bright ideas that adamantly refused to stay caught.

Schutzenberger paced the stuffy living room of his apartment, making a circuit from the green easy chair, with its torn slip cover, to the credenza and over to the windows and back again to the easy chair.

"Let us contemplate the following experiment," Schutzenberger said, "the Weizenbaum Experiment."

Joseph Weizenbaum of MIT had just published a book on the misuse of computer theory. Largely because *he* wished that he had said many of the things that Weizenbaum *had* said, Schutzenberger had taken an immense boisterous dislike to Weizenbaum and hence thought it appropriate that our experiment be named after him.

Now mention thus of an experiment may suggest that something or other is about to be executed or performed; in fact, the Weizenbaum Experiment is one of those purely imaginary affairs in which trains of thought are allowed to meander and then merge by anastomosis.

"In life," Schutzenberger went on thoughtfully, "there are two mathematical structures or spaces. There is a space made up of DNA or the proteins. This is an alphabetic space. Its objects are words. And there is a zoological space. Its objects are organisms. This is a space of alphabetic *representatives*. We say that in both spaces a natural metric exists – *the* natural metric – and that evolution proceeds in both spaces according to this natural metric. What is more, there is a mapping between the two spaces. It is this mapping that establishes that *my* DNA serves to express *me*.

I nodded and began to take notes.

"Furthermore," Schuztenberger said, "there is in the space of the nucleic acids or the proteins a probability transition system."

"This probability transition system – do you have in mind a finite-state Markov process?"

"Yes," said Schutzenberger dreamily, "a finite-state Markov process."

Schutzenberger stopped pacing and folded his hands in front of him, the long, curled, tobacco-stained fingers locked.

"We now observe," Schutzenberger said, "that the probability transition system is roughly in accord with the natural metric. We are speaking only of the alphabetic space, remember."

I looked up, for the moment unconvinced. "Why?"

"We say that the probability transition system is in accord with the natural metric *because* the most likely changes in the system are those that transform strings into *nearby* strings."

"I don't see this principle follows from the idea that alphabetic changes are independent."

"It does not," Schutzenberger agreed. "It follows from the observation that the probabilistic structure of the alphabetic space is not *uniform*. Indeed, it is this observation that shows ultimately that life does not comprise an ergodic system."

This was the sort of lovely lunatic leap that Schutzenberger was forever taking in conversation. I must have looked up with an expression of radiant confusion; Schutzenberger directed a warm, beaming smile into the ambient atmosphere, and went on, untroubled by my lack of confidence.

"Life," he said, "is conservative. Not everything that can change does change. For the most part, biological strings do not change at all. When they do change, they change in only one position. It is highly unlikely that a given string will change in respect to *every* position. We do not in life see a strand of DNA change its character at every possible codon in one sudden mutation."

I caught Schutzenberger's point.

"Now we need something more," he said, with the air of a man constructing a wonderful instrument, "a mirror, so to speak. We wish to say with regard to arbitrary strings not only how far apart they are under the natural metric, but how far apart their representatives in the real world are. For this we require an *induced* metric. Very common in mathematics."

I stopped writing to look up, and shook my hand to release its cramp.

"So when we talk of strings of DNA or strands of protein," Schutzenberger went on, "we can talk of the natural distance between them or their induced distance. Two sets of strings may be close under the natural metric and far apart under the induced metric. You know, there is the famous experiment in which chimpanzee and human polypeptides were compared. Simply considered as strings there is virtually no difference between them. Evidently there is some difference between a chimpanzee and a human being."

We were for the moment both quiet.

"In fact, zoologists often *assume* that the chimpanzee and the human being are closer than they really are." Schutzenberger held his own hand in the air, palm outward.

Having grown impatient with his own digressions, Schutzenberger finally said, "Let us now *perform* the Weizenbaum Experiment. We suppose that we have certain strings of alphabetic objects, and that there is some initial probability distribution defined upon them. That is to say, at the beginning of the experiment, the strings are most likely to be in a certain initial configuration. We also have – are you writing? Good – a probability transition system, one that tells which changes in the strings are probable and which are not."

Schutzenberger now resumed his studied pacing, moving from north to south in the stuffy living room, and then from south to north, the pale, pink light of late afternoon at his back.

"We select a distinguished element from the space of alphabetic representatives. This functions like a distinguished point in topology."

"Could we say," I asked, "that the first space consists of English letters, the second, of English words?"

"Of course," said Schutzenberger genially, "but the

genius of the Weizenbaum Experiment lies in its generality."

"I understand. But using letters and words as an example, can we say that letter-like sequences are close if they agree in spelling, and words, if they agree in meaning?"

"Exactly," said Schutzenberger. "We start, as I have said, with a distinguished element, a word, if you like, or an organism, or even a computer program. We consider now the set of all alphabetic sequences that are at an average distance from this distinguished element. Not too far away and not too close. We allow the alphabetic objects to undergo a series of transformations in accord with the laws governing their evolution. After a certain finite time has passed, the Weizenbaum Experiment has been completed."

"That's it?" I said. "That's the whole Weizenbaum Experiment?"

"It has the simplicity of great art," said Schutzenberger.

"What is this experiment supposed to tell us?"

"That depends," said Schutzenberger, "on whether the Weizenbaum Experiment has been a success."

In one of those curious flash-forwards I thought I saw the entire course of the discussion.

"We say [Schutzenberger used the royal and not the interpersonal "we"] that the Weizenbaum Experiment is successful if after a certain finite time, those strings that are at an average distance from the distinguished element come closer and closer to it."

"Let me get this straight," I said. "Suppose that our alphabetic space consists of letters; their representatives are words. Let's say our distinguished element is the word GODS. We consider sequences of letters that are at an average distance from this sequence, say within two alterations of the word: NIDS and KIDS but not NORDS and LORDS. We allow these sequences to change according to the probability transition system. If they begin to converge toward the word GODS, the Weizenbaum Experiment is a success."

Schutzenberger nodded. He then asked: "What can we conclude from the fact that a Weizenbaum Experiment *is* successful?"

I looked up, for the moment concluding nothing.

"We conclude, naturally enough, that the probability transition system *cannot* be arbitrary with respect to its *induced* metric structure. If words that are at an average distance from GODS move closer and closer to GODS, this must be because the probability transition system is *coupled* to the induced metric. Otherwise, there would be no reason for these words to move in this direction rather than any other."

"So a successful Weizenbaum Experiment demonstrates that the probability transition system is in accord both with its natural and its induced metric. This makes it seem very much as if a successful Weizenbaum Experiment were a contradiction in terms."

"Of course," Schutzenberger said. "In a certain sense, a successful Weizenbaum Experiment *is* an impossibility. It is also true that no one has ever seen or heard of a successful Weizenbaum Experiment."

I wrinkled my brows to signify perplexity.

"With one exception."

"And that is?"

"Life itself."

Eigen-values of Natural Selection

Evolution is a kind of majestic wave, moving ultimately from a primitive complexity to the dense, specific, integrated, miraculously complex order that is characteristic of, say, the human nervous system. In Darwinian thought, the effects of randomness are played off against what biologists often call the *constructive* effects of natural selection. In many respects, natural selection functions as a force-like concept, and, as such, acts locally if it acts at all. In speaking so sternly of locality, I mean to evoke the physicists' unhappiness with action at a distance. There is some controversy yet in theoretical biology concerning the ultimate unit of evolution. These I set to the side. I assume in what follows that evolutionary change takes place on the level of the nucleic acids and the proteins – within a purely typographic space. In this domain it follows that strings that are far apart under their natural metric should be weak in mutual influence. This is a space-like constraint. Then again, no string should be influenced by a string that does not exist. This is a time-like constraint, a rule against deferred success. It is a rule that appears often to be tacitly violated in evolutionary explanations. The historical development of a complex organ such as the mammalian ear involved obviously a very long sequence of precise changes. Comparative anatomy suggests that the reptilian jaw actually migrated earward over the course of evolution. It is very difficult to understand why each of a series of changes in the anatomy of the reptilian jaw should have resulted in a net increase in fitness. The rule against deferred success is violated when somehow the biologist points to the completed mammalian ear to provide a backward-looking explanation for changes in the reptilian jaw. This example may seem crude to the point of parody, but the rule against deferred success is often violated wholesale in theoretical biology when the examples become more sophisticated.

I have pictured evolution on the molecular level as a process involving paths. At any particular time, at any particular place, one has an ensemble of protein strings embedded, so to speak, in an underlying probabilistic structure. To this structure, natural selection is grafted, and acts, presumably, by measuring the degree to which a protein ensemble – a set of strings – is more or less fit. These measures I shall term the *Eigen-values* to the system. Suppose now that I consider a finite-state system consisting of an alphabet of 26 letters. The universe of strings drawn on this alphabet consists of all sequences k places in length. There are, of course, 26^k such sequences in all. I imagine the letters displayed on a linear array of some sort, a set of k squares, say. Each letter in the alphabet is assigned a fixed and independent probability of occurrence – 1 in 26 to make the example especially easy. An initial probability distribution fixes the configuration of the system for its first integral moment; at each subsequent step, every square changes randomly.

What are the chances, one might ask, that a system of this sort might converge on a particular sentence of English? Following Mannfred Eigen – he of the Nobel Prize – let us suppose that the sentence in question is TAKE ADVANTAGE OF MISTAKES. This is the *target sentence* – *S*; k is thus 27 (each space counts as a letter).

Even here, poised between irrelevance and imprecision, delicate and important biological questions arise. While it makes sense of sorts to say that for every string there exists a target – there would thus be many target sentences – it makes far less sense to say, as Eigen does, that there exists a target for every string, just one, in fact. Fixed in advance, a target so singular would seem suspiciously like a goal and hence *streng verboten* in evolutionary thought. How might such a target be represented, and by what means might its influence be transmitted to strings? These are not trivial questions.

In any event, nothing in Eigen's own example quite indicates why a stochastic system with a target sentence, however defined, should stop when it has reached its goal. This, however, is a trivial defect; it may be rectified by the construction of an *evaluation measure*. Suppose for the sake of simplicity that fitness involves only a mapping from strings to the numbers 0 and 1. At S – the target sentence – $f(S) = 1$; elsewhere, f is 0. An evaluation measure serves to size up strings in point of fitness as they appear: At S, it orders the system to stop; at all other strings, the command is simply to mush on. Stochastic device, target sentence, fitness function, and evaluation measure, taken as a quartet, make up an *Eigen system*. With the Eigen system in hand, let me return to my original question: How long would it take for a system so set up to converge on its target sentence? The enterprising Professor William R. Bennett, Jr., has actually performed the calculations. An arbitrary Eigen system would require virtually an infinite amount of time to reach even a simple target sentence. The figure that he reaches represents a number roughly a trillion times greater than the life of the universe. The problem plainly is not simply one of finding the right letters: There is also the matter of not losing them once they are found.

What more, then, is required? The opportunity, Stephen Jay Gould remarks, for the system to capitalize on its *partial* successes. Curiously enough, this is Eigen's answer as well, a bizarre example of independent origin and convergent confusion in the history of thought. As Eigen works through his examples, his system is designed to retain those random changes that fit the target sentence. Looking at the record of Eigen's own simulation, I observe that quite by chance the letter A appears in the first generation of strings, and at just the right place in the sequence – the second spot. It stays intact, A-ishly, for the rest of the simulation. When an E pops up in the fourth place on the string, it, too, gets glued in place.

An Eigen system with this sort of sticking power in place is an *advanced* Eigen system; it represents an obvious improvement over the hopelessly slow Eigen systems that I have already described. Under the advanced Eigen systems, fitness is no longer an all or nothing affair. f thus takes values, I suppose, *between* 0 and 1. Scanning every new string, the evaluation measure selects those strings S_i such that $f(S_i) > f(S_{i-1})$. These strings the system retains until it finds yet another string superior in fitness to the strings it already has. The result, of course, will be a sequence of strings that ascends steadily in fitness. At S, as before, the system stops.

There is no question but that an advanced Eigen system may reach an arbitrary target system in rather a short time. Unfortunately, in theoretical biology, as elsewhere, the question is not whether but *how*.

To the extent that fitness is a local property, it is difficult to understand why *every* ascending sequence should necessarily converge toward a neighborhood of 1, and hence indirectly toward S. A string that only partially conforms to S (TABB ADGANTIRE FO MIQRLMSO, for example) is locally no fitter than a string that remains resolutely unlike S. On the other hand, if each of the ascending sequences converges to S, it is very hard to see that fitness is a local property, and hard thus to understand what it is that an evaluation measure manages to measure. What is unacceptable from the point of view of theoretical biology is the idea that an evaluation measure judges fitness by calculating the *distance* between random strings and a target sentence. Distance is not a local property. An evaluation measure so constructed would plainly be responding to signals sent from the Beyond, a clear case of action at a distance and a gross violation of the rule against deferred success. In fact, this is precisely how the advanced Eigen system accomplishes its labors. An arbitrary string in which the letter A appears in second position is reckoned fit only to the extent that it is closer to the target sentence (TAKE ADVANTAGE OF MISTAKES) than it might otherwise be. The very concept of a target sentence thus constitutes a beery and uninvited guest in evolutionary thought.

Need I insist that the situation is made worse and not better if I allow the advanced Eigen system to stop when it has reached any sentence of English instead of a particular target sentence? I suppose so. The point appears to have confused a great many biologists. It should be plain in the general scheme of things that a target sentence is a minor stand-in for a major concept. Now if an advanced Eigen system is deprived of a particular target, then it must be designed to stop once it has reached *any* sentence of English. It must thus be provided with an abstract characterization of all the English sentences that may be drawn on a 26-letter alphabet. Of these, there are a great many. An Eigen system bouncing briskly from one set of random permutations to another, no less than the linguist or the logician, requires nothing less than a grammar of the English language if it is not to keep babbling forever. And the concept of a grammar plays no role whatsoever in contemporary evolutionary thought.

Florence

Once Schutzenberger arranged for the laboratory to travel to Florence for a conference at the Institute for Global Analysis. We left Paris in the late afternoon of a damp, drizzling day, our entire group of fifteen traveling together on the slow overnight train that leaves the Gare du Nord at around five in the afternoon and arrives the next morning in Florence a few hours after dawn. Everyone brought his own food. I found myself in a compartment with Maurice Gross and Schutzenberger. As the sun fell over the flat fields of southeastern France, Gross brought out a bottle of red wine, and, when that had gone, another, and yet another. The light acquired for a time an eerie fluorescent glow. Schutzenberger and Gross talked steadily, snickering and smiling, the French sentences like the long listing glide of an evening bird.

French railway tracks are seamless. To an American, used to a monotonous Amtrak clunk, the sound seems strange, a steady elegant hiss. At each station the sequence of steps, and the scene beyond, was the same. The long clumsy train would come to a halt with a great melancholy release of air from its steam brakes. From outside, a muted gabble of rapid-fire French. Looking through the smudged window, I could see a row of tall lamp-posts, their tops curving into a glowing, bug-spattered tube. On the station house wall, facing the tracks, was a perfectly round clock, with elegant, spidery hands, and a row of tattered and defaced advertising posters, chiefly for Kroenenbourg Beer or a horrible instant soup; beyond, there was the empty ticket room, with a single spare wooden bench. On the station platform itself, an old-fashioned luggage cart, its gap-toothed prongs dug into the concrete, stood posed companionably beside a very old woman – the same woman at each stop – who stood staring somberly into space. A whistle would blow and then our plump, energetic conductor would walk rapidly from the front of the train to the back, peering underneath the carriages, and finally swinging up the steps as the train moved from the station.

We arrived in Florence the next morning, groggy and hung-over. Everyone was happy to be in Italy. The light was strong, vibrant, bathing the far hills in a flamboyant series of browns, umbers, and faint rufous reds. We trooped off to our hotel, where, of course, the proprietor, standing before a greasy ledger in his undershirt, his palms held upward, explained in his pidgin French that he had never received our reservations. Without washing or changing our clothes, we headed toward the museums, and the piazzas, and the great churches, and those marvelous Florentine coffee shops where indescribably pretty young women gather to sip sickly-sweet drinks and exhibit their gold-coppery shoulders.

Our routine was always the same. Lectures at the Institute began at eight in the morning. Speaking no Italian beyond *Prego* and *Grazzia,* I delivered my lectures in English. No one understood a word of what I had to say, but everyone smiled broadly at the conclusion of my talks as if to suggest that in Italy intellectual communication was generally achieved wordlessly. We broke for lunch at one, taking our meals in a noisy *trattoria* presided over by an immense, garlic-smelling woman, who managed miraculously to put food on the table by means of a kitchen that so far as I could tell was no bigger than a foyer or a cabinet. At four, the working day was over; we spent the rest of the time sight-seeing. At ten or eleven in the evening, we met again for a meal that might last until one in the morning.

Schutzenberger was charming at lunch, but positively scintillating at dinner. All through the meal, he would talk in a steady, fascinating, mocking, derisive, and malicious way. He talked about art and about politics, and about science and about mathematics. He had lived in Southeast Asia. He had seen China during the revolution. He had been a physician. He knew everyone everywhere. He would talk about the architectural marvels of rural Cambodia, the expression on his face signifying that what he had seen had stretched to the limit his capacity for the assimilation of beauty, rendering him almost speechless – *fantastic,* he said sadly – and the correct treatment of Yaws. His voice was musical, his French, liquid and alive. He not only talked, he listened, always ready to be amused. He had a trick of smiling in anticipation. Before a story was complete, he would crook his narrow flushed face into a raised bird wing smile, his features separating into two symmetrical planes, the bad teeth showing broadly, the eyes pinched. By the time dessert was brought, we were all tired and punchy from too much wine and food.

By and by we walked home through the darkened

shabby streets of Florence. It was then almost three o'clock, but Schutzenberger was not ready to call it a night, and so he sank to the stoop of our hotel, and we sat on the curb in front of him, unwilling to let go, unable to keep up. A gray pallor of exhaustion had spread over his face. The lines of age that made him look older than he was had darkened and deepened. He sat smoking for a while, talking of this and that, commenting on nothing at all. Our attention flagged. From behind a parked car, a long, elegant, Florentine night cat appeared, and struck by the fact that here we were, on his street, in the dead of night, sidled over for a closer look. Schutzenberger saw him coming and turned to face the cat. Very slowly, in a low, vibrant voice, he began to talk, gently, soothingly. Transfixed, for no one had ever talked to him like that before, the cat sank to his haunches, licked each paw carefully to make sure they were clean, and then cocked his head toward Schutzenberger to listen more attentively. Schutzenberger talked to the cat about the evening, and the weather, and about Florence and its streets; he talked about cats that he had known and their sad destinies. He talked about the stars and the secret signs of the night sky, pointing with a raised finger to the heavens; he talked about dogs with a deep frown of distaste. He talked about the natural numbers; he talked about the end and the beginning, and much besides. The cat listened to all this, ears cocked, whiskers twitching, but then, finally, remembering that he had business to attend to elsewhere, stole softly away.

David Berlinski is a philosopher, writer, and translator. Since earning degrees in history from Columbia College and in philosophy from Stanford University, he has been Assistant Professor of Philosophy at Rutgers University, a Fellow of the Faculty in the Department of Mathematics at Columbia, has translated works from French, and has written screenplays and television scripts. Deeply interested in the history of science and ideas, Mr. Berlinski has lectured throughout the United States and Europe. The author now divides his time between San Francisco, Paris, and Vienna.